THE RAILWAYS

OF THE
MANCHESTER
SHIP CANAL

THE RAILWAYS
OF THE
MANCHESTER SHIP CANAL

Don Thorpe

Oxford Publishing Company

ISBN 0-86093-288-5

Typesetting by:
Aquarius Typesetting Services, New Milton, Hants.

Printed in Great Britain by:
Netherwood Dalton & Co. Ltd., Huddersfield, Yorks.

Published by:
Oxford Publishing Co.
Link House
West Street
POOLE, Dorset

Acknowledgements

This book is dedicated to all railwaymen but especially to the staff, past and present, of the Manchester Ship Canal Company.

For the help and encouragement given to me I thank the following:
Mr A. S. J. Campbell, Chief Mechanical & Electrical Engineer, Manchester Ship Canal; Mr B. Valentine, Railway Superintendent, Manchester Ship Canal; Mr B. Vaughton, Public Relations Officer, Manchester Ship Canal; Mr D. Thornley, Press Officer, Manchester Ship Canal.
Also members of staff of the aforementioned departments and many other departments.

Of the assistance received from sources outside the Company, each of which is acknowledged either under photographs or on this page, a special word of thanks is given to the Industrial Locomotive Society, and also to Geoff Holt and Tony Wadsworth for use of material written on the subject in their youth.

A heartfelt vote of thanks is given to Elaine Banister for her labour in typing from such a poor manuscript, and remaining to complete the course, and also to Sheila Barlow for her share in the typing.

Finally I thank my wife, Pat, for her exceptional store of patience and understanding, with so many household tasks going neglected, my brother, Bob, for his photographic advice and my mother for her encouragement.

Further Acknowledgements

Thanks are also due to Dr J. A. D. Ackroyd; Associated Octel Limited; Department of the Army Office of the Chief of Military History Washington, D. C.; M. Beeston; Bowater Limited; British Aluminium Company Limited; British Steel Corporation; British Tar Products Limited; K. Butterworth; C. J. Burns; Cabot Carbon (UK) Limited; Colgate Palmolive Company Limited; Containerbase Limited; Courtaulds Limited; CPC Limited; A. Day; J. Derek; S. Dolan; J. Drinkwater; Dupont (UK) Limited; Geigy Company Limited; A. Hayman; Dr. J. P. Hodgson; Hunslet (Holdings) Limited; I. C. I. (Dyestuffs) Ellesmere Port; I. C. I. (Mr Allen); S. Jackson; A. Jarvis; Miss L. Jones; W. W. Jones Dooly Limited; H. Joyce; Kellogg Company of Great Britain Limited; Lancashire Tar Distillers Limited; Laporte Industries Limited; G. Leach; C. Lever; J. Lloyd; J. Locke; Mrs Loddington; J. Lowe; Libraries at Eccles, Ellesmere Port, Irlam, Manchester Central, Salford, Stretford, Urmston, Warrington, Widnes; Sir Robert McAlpine; National Coal Board; R. Nichols; Mrs E. Oakes; Dr D. Owen; Pan Ocean Terminals Limited; H. Pennington; Port of London Authority; R. Potts; R. Prince; Proctor and Gamble Limited; R. N. Redman; F. Shawcross; Shell (UK) Limited; J. Shelden; Mrs J. Swindles; C. Thompson; F. Tonks; Trafford Park Estates Limited; Turners Asbestos Company; Vairens Limited; W. Wardley; S. Williamson; W. Williams; City of Chester Archivist; City of Salford Archivist;

Contents

Foreword

Five years of devoted work have gone into the making of this book. I say making, deliberately, because in those years it has involved the author in travelling extensively, sometimes at the most awkward hours, writing endless letters, interviewing, seeking information, even pleading for it, photographic work and, of course, hours of research. Indeed, the whole effort has been a labour of love, and Don Thorpe is to be congratulated on the result.

It is difficult to say in a few words why the railways hold such an attraction for so many people. From the youngest train spotter to the retired clergyman expertly photographing scenes of railway activity, the interest amongst all groups is world-wide.

The Manchester Ship Canal Railway was somewhat unique. The construction of the canal, during the peak of the railway age, was probably the biggest civil engineering feat of its day. The product of a proud and independent city which, seeking to solve its problems, decided to bring the sea to the city in order to consolidate its commercial prestige. Railways were an essential part of that development. Every quay, wharf, transit shed, warehouse and private trader, within the canal estate, was equipped to handle rail traffic. Over two hundred firms in Trafford Park had a private siding connected to the Manchester Ship Canal network. These were, in turn, through various junctions, linked with the railway companies. But facts and figures are only the skeleton. The life was provided by the thousands of men who, over the years, worked the railway; locomotive drivers and shunters working all sorts of peculiar shifts in all types of weather, and platelayers putting down first class track with the minimum of mechanical aids. Engineers kept the system going and a clerical and commercial organization ensured all went smoothly. To all these, I know, Don Thorpe has dedicated this book and for my part I am proud to be associated with it.

The late B. Valentine
Railway Superintendent
Dock Office
September 1981

Introduction

The Manchester Ship Canal railway system is well-known in industrial railway circles, and therefore it comes as something of a surprise to realize that so little has been written about it. That which has been published has appeared mainly in railway magazines and is necessarily in a condensed form. Nevertheless, these articles provided a valuable record of the railway at the time they were written.

However, in this account an attempt has been made to cover every aspect of the Manchester Ship Canal system, its beginnings, its growth, its role in times of war and peace and, of course, the people who were part of it.

At its zenith the Manchester Ship Canal Railway was the largest privately-owned railway in Britain, with a total of 75 locomotives, almost 2,700 wagons and 231 miles of track. It employed over 790 persons, the largest of any group working in one Manchester Ship Canal operation.

Today the railway operates on a very much reduced scale, due to the adverse effects of changes in the economic and industrial climate coupled with the ever growing competition from road transport. Whilst certain factors which part of the overall picture; a great deal of information, unfortunately, not being readily available. Even the Company's own records concerning its railway system have, of necessity, been reduced over the years. Much was lost during World War II and later, lack of space determined what could be saved.

There are many myths and misconceptions surrounding the Manchester Ship Canal Railway, and some of these are discussed and, in a few cases, clarified. However, even though events have been put into a proper perspective, the truth has not detracted from the story but has, to some extent, added to it.

It has always been stated that the railway system was based on the lines of the contractor's system but, for the major part, this was not so; there being very little of the contractor's track left in those sections where it would have been of use to the infant Railway Department. Many people refer to the system as 'the docks railway', but in reality it provided a service not only for the docks but also for a large number of industrial premises. On most maps the track is shown as a 'mineral railway', but whatever the cartographers termed it, the Manchester Ship Canal Railway was never simply that. True, minerals were carried; coal, ore, sulphur, and others, but there was also cattle, sheep, steel, wheat, timber, coach bodies, cars, fruit and machinery and so much more.

The title of the book may seem ambiguous but the facts are, that apart from being split into several isolated sections, the railway was operated as two entirely separate entities; it was also developed, over a period of thirty years, using earlier existing railway systems.

Over the last eighty years there have been many works published concerning the Manchester Ship Canal and itself and its raison d'être, therefore, although it is useful to include details of events leading up to the start of the construction of the Canal, it is not proposed to dwell on these events to any great length.

Pre-History of The Canal

Briefly, the idea of building a ship canal had been mooted as far back as 1824 because the passage of water-borne traffic using the River Mersey and River Irwell was too slow and uncertain, also like that on the Bridgewater Canal, was of too small a capacity to cope with the needs of an industrial city like Manchester. With the opening of the Liverpool & Manchester Railway in 1830, the prospect of swifter, cheaper transport was envisaged by the people of Manchester, but after some years of operation it was seen that swifter it may have been but cheaper it was not. Due to the greedy men of the railway companies and of the Port of Liverpool, costs were outrageously high. The actual charge on goods shipped across the Atlantic Ocean was many times lower than the levies of the Port of Liverpool or the railway charges between that city and Manchester.

During the latter part of the 19th century, Manchester was seen as a declining city with empty properties in abundance, and with industries and workers moving to other parts of the country to escape the burden of costly transport, with which the area was saddled.

The battle to reverse the trend really commenced with the historic meeting, on 27th June 1882, between a group of northern businessmen at the 'Towers', Didsbury, the home of Mr Daniel Adamson. It was at Daniel Adamson's request that the meeting was held, and its purpose was to discuss the building of a ship canal. Needless to say the proposals were accepted in principle, and the movement was launched to obtain the necessary Act of Parliament.

It was to be over five years before the many obstacles had been surmounted and the work of actually digging the canal started. The fight to get a successful Bill, the vicious campaigns waged by certain railway companies, the various Liverpool interests and others, the bitterness created at the time are not dwelt upon in this narrative. It is sufficient to say that there were three attempts by the Manchester men before they were successful in gaining their Act of Parliament.

The first Bill was deposited for the Parliamentary Session 1833 on 17th November 1882. It passed through the House of Commons on 6th July 1883 but was rejected by the House of Lords on 9th August 1883. A second Bill was deposited on 21st December 1883 but although it successfully passed through the House of Lords on 24th May 1884, it failed on 1st August to get through the Commons. The third, and final, Bill was deposited on 16th December 1884, passed through the Lords on 7th May 1885 and through the Commons on 3rd August. Royal Assent was received on 6th August 1885. Incorporated in the final Act was the Manchester Ship Canal Railway Company.

The Provisional Committee had spent £172,000 in obtaining the Act, and they had also battled for over three years, but the final design was much better for the delay. Mr E. Leader Williams, who was responsible for much of the design, was appointed Engineer in Chief whilst Daniel Adamson, whose drive and enthusiasm had kept the whole project going, was made Chairman of the Provisional Committee.

Chapter One
The Construction Period (1887-1894)

The task of obtaining the Act had cost the Manchester Ship Canal Company a considerable amount of money, but once it had received the Royal Assent there came the even more formidable task of raising the capital necessary to build the Canal. At that time this was estimated at £6,000,000. A further two years was to elapse before they were able to start the work of actually digging 'the ditch', as its many opponents were calling it.

The largest single item on the expense sheet appeared before the contractor had even moved a spade on the site, for on 3rd August 1887, the Company purchased the Bridgewater Canal, its terminal docks at Manchester and Runcorn, and Duke's Dock Liverpool. The price paid was £1,710,000, and this was reputed to have been the largest amount paid on one cheque up to that date. Although the Bridgewater Navigation Company or, to give it its proper title, the Bridgewater Navigation & Mersey & Irwell Undertaking, was primarily concerned with the Bridgewater, a barge canal, the principle shareholders were, in fact, railway people. A more detailed account of the Bridgewater section and its history is given in Chapter 2.

The contract to construct the Ship Canal was awarded to Mr T. A. Walker, late of the Severn Tunnel, Barry Docks, Preston Docks, and many other major building projects. Thomas Walker was a remarkable man and a protégé of that even more remarkable man, Mr Thomas Brassey, often called the 'King of Contractors', a man whose record of building railways throughout the world has never been surpassed.

Mr Walker, like Mr Brassey, was a meticulous and humane employer. They both shared the same concern for the men working under them and provided not only for their bodily needs but also for their spiritual needs. Apart from a certain amount of living accommodation, there were hospitals, churches and schools for the workers' children.

Thomas A. Walker also believed in mechanization and that included the provision of a complete railway system for each section of the site. This was his main priority after the first cut had been made at Eastham on 11th November 1887. The rail used was 56lb. flat-bottomed type, and the cost averaged out at £630 per mile of single track. It was reported that the site railway was laid from Eastham to Manchester, a distance of 36 miles. In the light of other reports this statement is open to argument and is discussed in more detail in the course of this chapter.

The contract was divided into nine sections, each with its own plant, staff and workshops. At Norton, in that first month, there were 1,000 men. By February the following year, the section covering Eastham to Ellesmere Port employed 800 men and 8 locomotives. In September, the overall figure for locomotives had reached 87 and that for wagons 3,000.

With a contract of that size, the largest to have been awarded in this country up to that date, a great deal more plant was needed than the contractor could supply. As his other contracts were completed he transferred the machinery to the canal site. To augment this, new equipment was ordered from the makers. In August of the first year, Mr Walker gave an order to The Ashbury Railway Carriage Company, who contracted to deliver 100 wagons each week for six months. Many of these were of a tipping type and were of small capacity. The design is, even today, sometimes referred to as a Manchester Ship Canal type, and they had the letters T. A. W. painted on the sides.

Orders were also placed with the engine builders for locomotives, the minimum number being at least 102. The lion's share went to Manning Wardle & Company, the firm supplying 40 locomotives. The Hunslet Engine Company, supplied at least 22 and Hudswell Clarke & Company, about the same. A list of all locomotives on the contract, including some that have not been verified, and also those bought by the Company for their Railway Department, is included in the Appendix.

The original contractor, as was stated earlier in the chapter, had built many other works and this was evident in the names on the locomotives he had brought with him when he started work on the Canal; *Cadoxton, Elham, Pontypridd, Severn, Douglas, Lancaster* all being named after places in the area surrounding the sites of his contracts. In typical fashion he had named his new locomotives in the same way. There was *Cornbrook, Sale, Widnes, Salford, Bolton, Weaste, Urmston* and many more. Some of these were photographed at the time and a few are included here on *pages 8 & 9*. Two military-named engines, *General Stewart* and *General Havelock*, were bought new by the sub-contracting firm of Monk & Newall of Brindle Heath, Salford, and were sold on completion of the Canal to a railway contractor, Mr T. Wrigley, whose name occurs again later in the story. The locomotive, *Lord Warden*, is a bit of a mystery, at one time being described as an 0-4-0 saddle tank and at another as an 0-6-0 engine. Each of the contractor's engines was fitted with downward projecting blocks of wood on the buffer beams, which were put there to catch the buffer planks of the small contractors' wagons and were usually referred to as 'dumb buffers'.

The Contractor's Railway

The railway was laid on both banks of the canal cutting wherever possible, some of it being perched upon trestling and dykes, but the largest proportion was, of necessity, laid on the bottom of the excavation. These lines were of a truly temporary nature, for they were constantly being moved as the cutting became deeper. Sleepers were often little more than sections of tree trunks, cut to suitable lengths, then sliced from end to end, making a half round section. Even so, on 22nd August 1980, a journalist on the staff of the eminent professional journal, *The Engineer*, after a visit to the site, reported that, the railway was well laid, and fished at the joints. One section of the route which was perched on trestles, was that which took the line from the west side of the stone-built Barton Road bridge to the east side of the original Barton Aqueduct, (Brindley's 'Castles in the air'), which was also built of stone, and which carried the Bridgewater Canal across the River Irwell. The contractor had intended that the railway should run under the road and the Bridgewater Canal (the latter already having a tunnel passing under it which carried road traffic). Unfortunately the tunnel tended to flood so the alternative was to run the line under the arches of the two bridges.

At this point it would be of interest to recall the earlier statement that the railway was laid from Eastham to Manchester, another statement being that it ran from end to end, and continued to say the Mail Train ran the full length

Piling at Ellesmere Port almost ready for the infill of boulder clay which, when consolidated, formed the embankment between the River Mersey and the Canal.

M S C Co.

A German land dredger, built by Lubecker Maschinenbau-Gesellschaft of Lubeck (usually referred to as the 'Lubecker' or simply 'The German') working on No. 3 section in 1888. The name of the locomotive heading the spoil train is unknown.

M S C Co.

Locomotive *Ince* with Mr C. Gregson and others, possibly at Runcorn. Mr Gregson was one of T. A. Walker's agents on the contract. The engine was involved in an accident at Norton Depot when it struck a points boy who was the son of the driver.

M S C Co.

Hunslet engine *Urmston* worked on the Manchester Ship Canal contract from new in 1888 and was retained after the Canal's completion, but was sold, circa 1898, to S. Pearson & Son Ltd. of Wootton Bassett.

Courtesy J. Steggles

Locomotive *Sankey*, possibly photographed on Frodsham Marsh in 1890. The engine was delivered new to Ellesmere Port in 1888. Note the crenellated chimney cap.

M S C Co.

Originally locomotive No. 2 of Messrs G. B.Godfrey of Hull in 1881, *Sale* worked on the Canal contract until being sold to Stanley Bros. Ltd., at Stockinford, where it was renamed once more, becoming *Stockinford*.

Courtesy A. L. Baxby

Contractor's platelayers and newly-laid track, as seen on 2nd May 1890. In the middle distance is a partly-assembled land dredger built by M. Boulet of Paris. To the right of the picture is Calamanco Lock and the River Irwell.

M S C Co.

The contractor's line, laid on a platform mounted on pilings and driven into the bed of the River Irwell. The masonry in the foreground is that of Brindley's Barton Aqueduct whilst the bridge, viewed through the arch, is the original road bridge.

Eccles Library

of the Canal each day using the overland route. However, in the 22nd August 1890 issue of *The Engineer,* it stated that the railway was laid on the south side of the Canal from Eastham to the westerly shore of the River Weaver, the River Weaver, about 500 yards across at that point, not being bridged. Even so, this does not mean that there was no line through the section, it could have been carried on the embankment or the piling in the river, the gaps in the embankment, which allowed shipping to pass through, possibly being spanned by either temporary swing or lifting bridges.

The path of the Canal followed roughly the general line of the upper part of the River Mersey as far as Irlam, and did the same from Irlam to Manchester with the River Irwell. Obviously, the Canal took a straighter route than the rivers, a fact that made the railway on top of the bank more expensive than normal because of the number of bridges involved. To cross the cutting and the rivers that meandered back and forth across the site, a total of seventy four temporary bridges had to be built. Some of them were quite long and lay askew to the line of the river at the point at which it was being spanned.

At the earliest possible moment the contractor connected his lines to those of the railway companies and, although they were no friends of the Manchester Ship Canal Company, when it came to making the connection there was not a lot of delay. The terms of their various Acts precluded them from refusing to connect to anyone with a reasonable purpose and, of course, whether the Canal was completed or not, there was a great deal of traffic to be handled during construction. The tonnage with regard to plant, equipment and supplies was enormous. At one time the monthly consumption of coal worked out at 10,000 tons and for cement the figure was 8,000 tons.

One of the many lightweight swing bridges built to carry the contractor's railway over sections of the excavation. Perched on the bridge is *Levenshulme,* Manning Wardle No. 1003, built in 1889 and delivered new to Ellesmere Port.

M S C Co.

Frodsham Score Cutting, photographed on 13th May 1890. On the opposite side of the embankment, in the right of the picture, is the wide expanse of the River Mersey.

M S C Co.

A Sharp Stewart engine *Duke of Normandy* at Acton Grange in September 1890. This locomotive was used as an ambulance in emergencies. With its sister engine, *Haro Haro*, both ex-Jersey Tramways engines, it was one of the two 2-4-0 engines on the contract and also the oldest, being built in 1860.

M S C Co.

By January 1888, the contractor had made a connection with the Cheshire Lines Railway, at a point just east of Irlam Station. There were connections to the LNWR at Latchford and at Runcorn. The docks at Runcorn were, in fact, connected seventeen years or so before the Canal was even planned, but that is another story and will be related in Chapter 2. A connection had also been formed with the LNWR/GWR Joint line, at Ellesmere Port Station.

By far the most difficult section in which the contractor chose to make a connection was that at Cornbrook, Manchester, this being made to the Manchester-Liverpool line of the Cheshire Lines Railway. The major complication was the Bridgewater Canal which lay between the two points to be connected. A bird's-eye view of the area would have shown the lines of the Cheshire Lines Committee and the Bridgewater Canal separated by just a few yards, whilst to the north of the Canal, and at a lower level, lay the Manchester Ship Canal (in the period under discussion, still the River Irwell). Unfortunately no details have been unearthed concerning the route or the point at which the contractor made the connection. With the limited space available it is difficult to visualize just how it was accomplished, unless it was taken across the obstacles on a viaduct, as did the permanent Cheshire Lines Committee sidings which curved first south then back in a northerly direction met up with the viaduct. At the docks end of the viaduct the line then curved westwards and descended, rather sharply it has been said, on to the side of the Ship Canal, passing the front of the Dock Offices before back in a northerly direction met up with the viaduct. At the docks end of the viaduct the line then curved westwards and descended, rather sharply it has been said, on to the side of the Ship Canal, passing the front of the Dock Offices before passing under Trafford Road. In the construction period the contractor's line crossed the River Irwell on a line perched on piling, but when the Canal was completed the line was taken across on a temporary single line swing bridge. The temporary swing bridge was quite a feat in itself; from ordering, the completion took a mere two months. It was a light steel structure with a length of 50ft. which pivoted on a roller ring mounted on a timber support.

Accidents & Collisions

Construction sites are never the safest places to be, even today, but in those days, with less safety regulations in force, major accidents were all too frequent. One terrible accident happened, in July 1981, when a loaded train of spoil wagons was being propelled to the top of the bank by two locomotives, but was switched, by mistake, into a disused spur line which ended on the brink of the cutting. There were no stops on the spur line and the train toppled over into the cutting. Below dozens of men were working and with the speed at which events happened, they had no time to escape. Ten men were killed and many more badly injured, and both locomotives were completely destroyed. Another less terrible accident, involving a father and son, happened at Norton Depot during the first month of the Canal contract. The son,

A contractor's signal box and signals at Mode Wheel on 21st February 1891. Note the extreme youth of the signal operator and the points boy. The warning notices on the cabin indicate 'No loafers admitted'. Behind the signals stands a travelling overhead crane for handling the stone blocks used in the construction of the lock sides and the quays. *M S C Co.*

A view of the cutting, as seen on 22nd April 1893, looking westwards towards Moore Lane swing bridge from the bridge of Railway Deviation Nos. 1 & 2. At this section, the sides were mainly of red sandstone topped with a layer of soil. Note the basket weave on the sloping soil layer and the track descending into the cutting. Traces of the ramps which were cut in sandstone are still visible in certain sections. *M S C Co.*

who was a pointsboy, was struck by an engine driven by his father, who was so badly shocked that he had to have medical treatment along with his son.

Collisions also occurred and, in some cases these were due to single line working. When the locomotives *Garston* and *Salford* collided there were no casualties, but when *Deal* and *Rhymney* collided three men were killed. *Deal* was taking the mail train when it happened and was moving quite fast; the line supposedly being kept clear for the train. It happened at the end of June 1890 and at the inquest, held in September of that year, the Coroner was told by one of the drivers that the mail was supposed to run through the section between 5.30 and 6 o'clock morning and evening. However, he complained, the rules were always being changed.

The crew of the *Lancaster* were more fortunate, when their engine was covered with water during one of the sudden floods which caused so much damage and delay to the cutting of the Canal. They just managed to scramble clear in time.

One series of incidents involving the railway was less tragic and, since nobody was injured, in retrospect, seems quite funny. This was a public relations exercise, the sort that P. R. officers the world over have nightmares about. It happened in June 1889 and involved a group of London journalists who had been invited by the Company to view the progress of the work. A trip by rail had been organized, the train being made up of a saloon coach and several open wagons. The fortunate people in the coach were all right, but the unfortunate ones in the wagons spent most of the time brushing burning embers, which rained down from the locomotive's chimney, off their clothes. To add to their misfortunes, the train brushed against a projection on one of the temporary bridges, bringing down part of the structure on to the wagons. The shaken journalists were none the worse for their experiences and all returned, quite literally, to tell the tale. However, public relations exercises under happier circumstances were more usual.

The Manchester Ship Canal Company become Contractors

In November 1891 the contractor Thomas A. Walker died, and the whole operation almost collapsed. In any event, the Company decided to make itself main contractor and work to complete the Canal. Before Mr Walker's death the weather had been reasonable and progress had been good, facts that probably swayed the directors in their decision to complete the Canal themselves. The death of Mr Walker started a long period of litigation between the Company and the executors of the late contractor. This dragged on for over a year until March 1891 when the Manchester Ship Canal Company agreed to pay the sum of £1,053,600 for the plant and equipment. This included 4 river dredgers, 96 steam navvies,

Contractor's equipment and horses in the Ince to Ellesmere Port cutting photographed, possibly, around 1890. In the background is a steam excavator, most probably built by Messrs Priestman & Co., whilst in the foreground is Hunslet 0-4-0 *Dallam* with its crew. Built in 1888, the locomotive, (Works No. 453), later went to T. Oliver, Contractor, and survived until after 1945.

M S C Co.

The cutting at Latchford, looking westward towards the foundations at Latchford Locks. The photograph was taken on 5th May 1890 from Thelwall Road which eventually was cut away. It is pictured before the embankment carrying the deviated lines of the LNWR had been started. Note the wooden chutes to the left of the picture which were used to convey building materials from the side-tip wagons, above, to the wagons in the cutting.

M S C Co.

Building the sluices at Weston Point on 10th September 1889. The engine on the timber-built viaduct is *Dynas*, a Hunslet-built engine which earlier worked on the Barry Docks contract, hence the name.

M S C Co.

192 steam cranes, 182 portable and other engines, 212 steam pumps and, of course, 173 locomotives, 6,300 wagons and 228 miles of temporary railway.

To digress, the figures given above are often quoted as the total for plant and machinery used in the construction of the Canal. In fact there was a lot more of it than that. Although the highest total of locomotives reported to have been present on the site at any period during the seven years it took to complete the Canal was 175, it is thought that the overall figure was probably in the region of 180 or even more (*see Appendix*). By the end of July 1890 there were 171 locomotives, 6,300 wagons and 221 miles of track. The figure by the end of August had increased to 175 locomotives and 6,500 wagons, but the track had decreased to 220 miles. There were 70 locomotives, 53 miles of track and 5,000 men working on the Runcorn-Latchford section alone, prior to its completion.

Deterioration of Weather

Apart from the legal wrangling which followed the death of Mr Walker the Company was beset by other troubles; the weather deteriorated and the area was swept by the worst storms experienced for many years. The site was flooded in numerous places several times due to torrential rain; gales lashed the embankment between the River Mersey and the Canal causing extensive damage to the lock walls near Runcorn.

Locomotives, railways, wagons and all types of plant and machinery were covered by several feet of water on more than one occasion, and this was followed by intensely cold weather which caused the floodwaters to freeze. All this caused the costs to rocket, the original estimate of £6,000,000 being seen to be totally inadequate. On a happier note, on 29th August 1889, Lord Tatton, addressing the half-yearly shareholders' meeting, stated that the Railway Companies were now willing to take their lines to the docks. This was a turn-round, since some of them had stated that they could not see the 'big ditch' ever being anything but a barge canal, that is, if it was completed.

Changes in Plan

One of the obstacles in the path of the Manchester Ship Canal Company's plans up until that date, had been the refusal by Sir Humphrey de Trafford to sell any of his land

The embankment under Runcorn Railway Bridge. This was the narrowest section of the Canal, the width being dictated by the bridge support foundations.

M S C Co.

Contractor's engine *Raglan*, in an unspecified section of the cutting. Delivered new to T. A. Walker on the Barry Docks contract in 1887, the Hunslet engine was then transferred to the Canal contract. On completion of the contract, *Raglan* was retained by the Company and, in 1915, became MSC No. 3 and was based at Mode Wheel Shed.

M S C Co.

Barton Bridge and behind this, the aqueduct with the masonry of the partly-demolished original bridge and aqueduct. The contractor's line in the left of the picture was the second to be put through the section, the first being laid on the opposite side of the cutting.

M S C Co.

to the Company, the section from Mode Wheel Locks to Trafford Road, Salford, being of prime importance since it made up one side of the 'pool' from which the docks were entered. However, in 1889 the younger Sir Humphrey de Trafford, following the death of his father, decided to sell part of his land. This now forms the frontage to the docks, Trafford Wharf occupying most of the length. Of course, this meant changes in the earlier accepted plan, a plan that did not include the railways on Trafford Wharf nor the bridge leading to that railway. Therefore, the Company had to seek Parliamentary permission to bring about such changes.

Railway Deviations and Swing Bridges

Referring to the period when the Company was attempting to put its earlier unsuccessful Bills through Parliament, amongst the proposals which aroused most opposition from the Railway Companies was that to take the existing lines, which crossed the path of the intended Canal, over on swing bridges or lifting bridges. Roads also were to be dealt with in the same way. Although the proposal for the roads was accepted, in the case of the railways it was unsuccessful, and rightly so. That would have been a retrograde step as far as the Railway Companies and their customers were concerned. The obvious solution was to take them over the Canal on high level bridges, building them to the same height as the Runcorn Bridge spanning the River Mersey, one which had carried the lines of the LNWR since 1868. The headroom was fixed at a minimum height of 75ft., which gave the bridges an approach gradient of no more than 1 in 135.

The routes that had to be dealt with in this way were labelled Railway Deviations Nos. 1 to 5. Nos. 1 and 2 were taken across on a single bridge, No. 1 being the LNWR Warrington to Crewe line, whilst No. 2 was the GWR Warrington to Chester line. Deviation No. 3 was the LNWR Warrington to Stockport line at Latchford. Both No. 4 and No. 5 involved the lines of the Cheshire Lines Committee, No. 5 being the Warrington to Manchester route, the new bridge crossing the Canal at Irlam. The new Irlam Station had to be raised to a level much higher than the previous station and was built on the opposite side of the old line which remained to pass to the rear of the building. At Glazebrook Junction, the lines spurred off the route just mentioned, eventually reaching Stockport. This became Deviation No. 4, crossing the Canal at Cadishead.

In order that normal working could be continued by the respective railway companies, the embankments erected to carry the new lines were located to one side of the original lines, which lay across the path of the Canal on their undisturbed banks of earth. Unlike the other Deviations, No. 3 was rerouted to the east side of the old lines. The steelwork for the high level bridges was supplied by the Glasgow firm of Sir William Arrol & Company. There was nearly a fifth high level railway bridge across the Canal, which was to have carried the metals of the proposed East & West Coast Railway. The new route was to have connected Warrington with Sutton on Sea on the Lincolnshire coast, and to have crossed the Canal at a point between Latchford and Wilderspool. In fact it did not get beyond the proposal stage.

Obstructions by Railway Companies

On completion of the high level railway bridges, the railway companies, in their usual obstructive way, refused to use them for anything but goods traffic, even after the Board of Trade Inspector had tested them and declared them safe. The test, by the way, being to set rows of locomotives with tenders across the whole length of the bridge, then check the deflections in the structure. This refusal to use the bridges for passenger traffic meant, of course, that both the old and new routes were being used by the railway companies, and naturally this prevented the removal of the soil banks which separated the completed, but unconnected sections of the Canal. The LNWR, the most bitter of the Canal's opponents, wanted 6 months usage of the bridges before they would be willing to use them for passenger work. However, the eventual delay was less than that, due to public agitation.

Birth of a New Port: Saltport

In the latter years of the contract, another port appeared on the Canal and it had a railway connection to the main line, that is the GWR/LNWR Joint line. The first mention of the possibility of a port facility in that area was made in *The Engineer* in an article issued early in 1891. It commented on the opportunity to use the stretch of water close by the mouth of the River Weaver, referred to as Weaver Pool, to load and unload ships. Whether the idea had already been considered by the Manchester Ship Canal Company is not clear, but by 6th November of that year it was being used for that purpose. The following year saw the new port facility in full swing, handling cargoes of pitchpine, grain, pottery, pig iron and salt. This part of Cheshire is, of course, a salt-producing area; no doubt this is what led to the name of Saltport being officially used from 22nd July 1892.

Railway Companies Act: Docks Connections

In the early part of 1890 both the Cheshire Lines Committee and the L&YR had petitioned for changes to their Acts which would allow them to construct railways to the docks at Salford and Manchester. The Cheshire Lines Committee was to build its junction at Cornbrook, close by the Cheshire Lines Committee-contractor's temporary connection which has already been discussed in some detail. The new Bill was granted on 24th April 1890.

The L&YR Act, granted on 4th August 1890, was for a double track line branching off the Bolton-Manchester Railway at Windsor Bridge and continuing to an area between 8 Dock and New Barns where, at that time, the Manchester Racecourse was located. There was also provision for another 250 yard extension which spurred off through a short tunnel southwards, ending up alongside the Trafford Road Dock wall in Salford Yard. The extension was later abandoned in the 1894 Act. Obviously, Salford Council objected to the Bill due to the number of houses that were to be affected. However, the Act was granted as it was originally proposed. The detail of the link is given in Chapter 2.

In February 1892 the LNWR deposited its Bill, petitioning for a connection at Weaste. This was to start at a point close to Eccles Station by descending a gradient then turning southwards to pass, first, under their Manchester-Liverpool route, formerly part of the original Liverpool & Manchester Railway and, second, under Eccles New Road, now the A57. The lines emerged between an area known as Little Bolton, and the sewerage works of Salford Corporation. A little way beyond this point, the LNWR lines met those of the Manchester Ship Canal.

Completion of Railway Deviations

The Railway Deviations were completed and taking goods traffic before the end of February 1893, the first being No. 5

Looking across the bridge of Railway Deviation Nos. 1 & 2 at Acton Grange. The bridge is shared by the lines of the LNWR and GWR.

M S C Co.

Mr Price's soil transporter with dredger, both working downstream of Warburton Bridge. The wagons are being loaded direct from the dredger instead of the more usual method of loading into barges. Another rather unusual combination of railway and dredging operations was the practice, in some sections, of towing the spoil barges by locomotives instead of tugs. The process was apparently less time-consuming! Although the locomotive in the picture is not specified, the maker was Manning Wardle and the style suggests *Bolton*, later MSC No. 7.

M S C Co.

which opened on 9th January. Nos. 1, 2, 3 and 4, opened on 27th February. The Cheshire Lines Committee, a group of companies whose opposition to the Manchester Ship Canal had not been quite so fierce as that of the LNWR, used their bridge at Irlam, (that is Deviation No. 5,) to carry goods for a mere ten weeks before taking passengers across it. The opening date was 27th March. Their other bridge, Deviation No. 4, was only restricted a short while longer, just over three months, before the passenger traffic started to cross the Canal on 29th May. True to form, the LNWR restriction on passenger traffic was not lifted for almost four months and that was only done under pressure after the Manchester Ship Canal Company had made strong protests.

Within a short space of time the contractor's machinery was eating away the earth dams that had carried the LNWR metals. The old lines were to remain connected to the railway companies' metals, being used for future junctions between them and those of the Railways of the Manchester Ship Canal Company. Towards the final stages of the construction of the Canal the necessity for such large amounts of equipment lessened. Auctions were held at various depots, the Company having already decided what plant to retain. The London firm of auctioneers, Messrs Fuller, Horsey & Company, held an auction at Salford Docks offering 200 lots. There were 14 large locomotives included, the best prices realized being £540 for one locomotive whereas a portable engine realized £180.

The Manchester Ship Canal had been built to overcome the 'bogey man' of industry, costly transport. The railway of the Manchester Ship Canal was merely an ancillary, but it was to play quite a large part in the development of the Company's land as an industrial estate. One of the earliest firms to be established along the banks of the Canal was that of Richard Evans, Haydock Collierys. Their wharf, at Acton Grange near Warrington, was completed well in time for the opening date, but the flood experienced on the last occasion caused extensive damage to it. It was repaired eventually, but it was rough luck on a company who had strived to be ready for the great day.

Along the length of the Canal everything was being cleared in readiness for the opening day on 1st January 1894. Staging and scaffolding had been removed, rough ground had been levelled, and most of the temporary railway lines taken up. The mammoth, and at times, seemingly impossible task was now at an end. A quiteness reigned on the scene that had witnessed so many heartbreaking set-backs and hard labour. As winter set in and the opening day drew near, most of the men who had worked on the site would be looking back over the seven years it took to build the 'big ditch' and be proud to have taken part in it. For those who had not worked on the project but had lived near to it, life would never be quite the same again. In those days, country people were used to retiring to bed soon after darkness fell, and to witness the sights and sounds of a work-force toiling night and day, must have been a spectacle that would remain in their minds eye for many years.

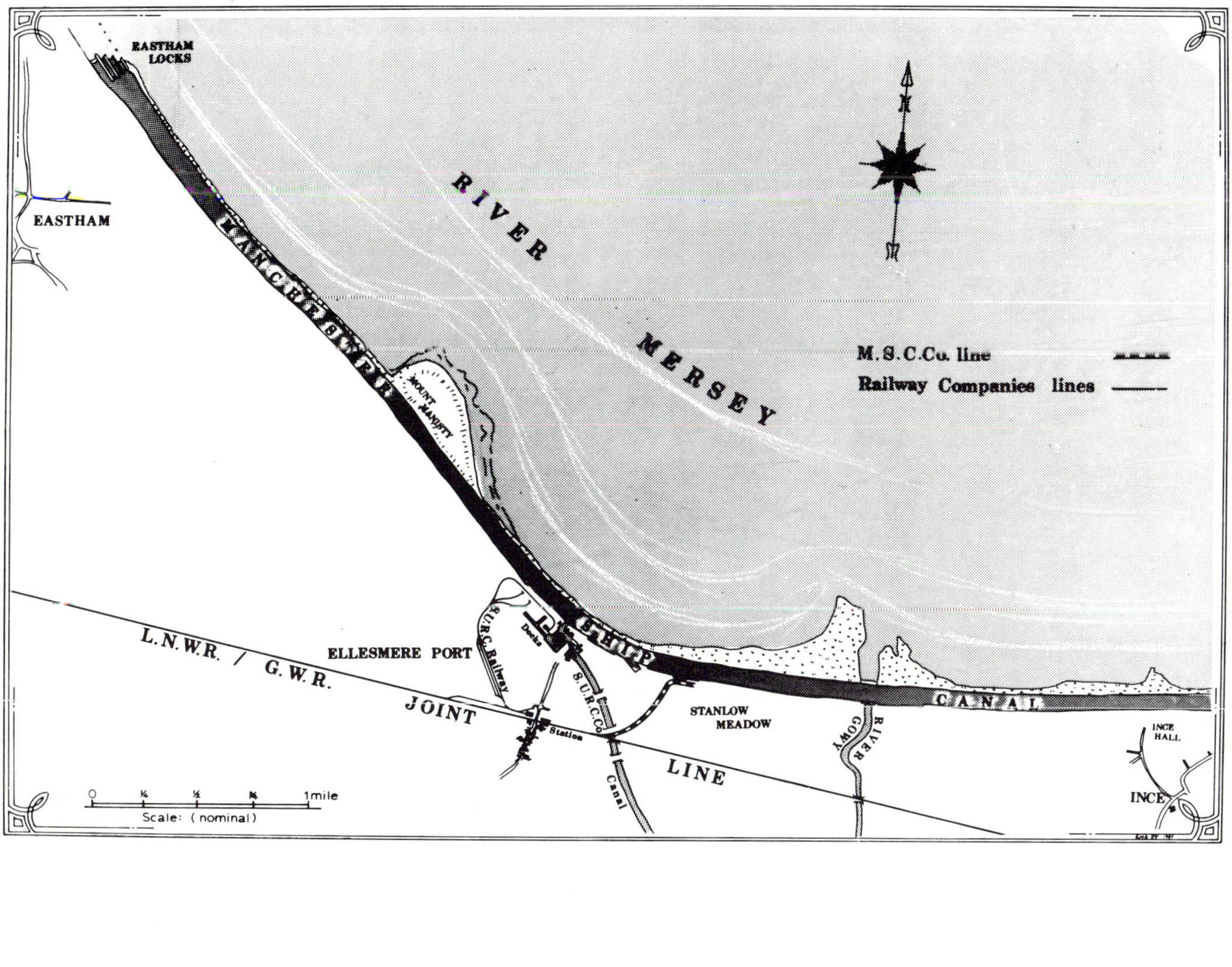

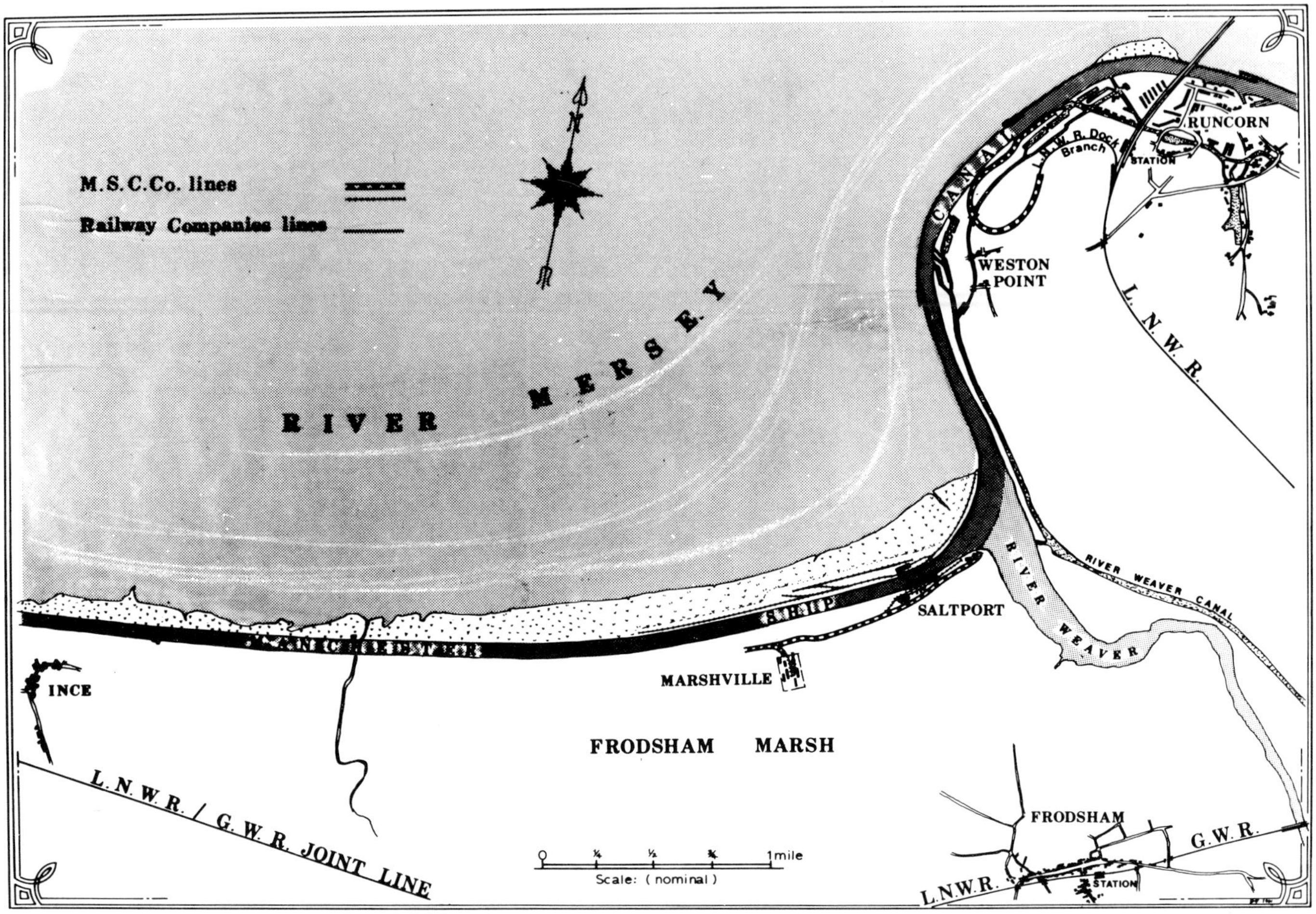
M.S.C.Co. lines
Railway Companies lines
N
RIVER MERSEY
CANAL
SHIP
RUNCORN
L.N.W.R. Dock Branch
STATION
WESTON POINT
L.N.W.R.
RIVER WEAVER CANAL
RIVER WEAVER
SALTPORT
MARSHVILLE
INCE
FRODSHAM MARSH
L.N.W.R. / G.W.R. JOINT LINE
FRODSHAM
G.W.R.
L.N.W.R.
STATION
0 ¼ ½ ¾ 1mile
Scale: (nominal)

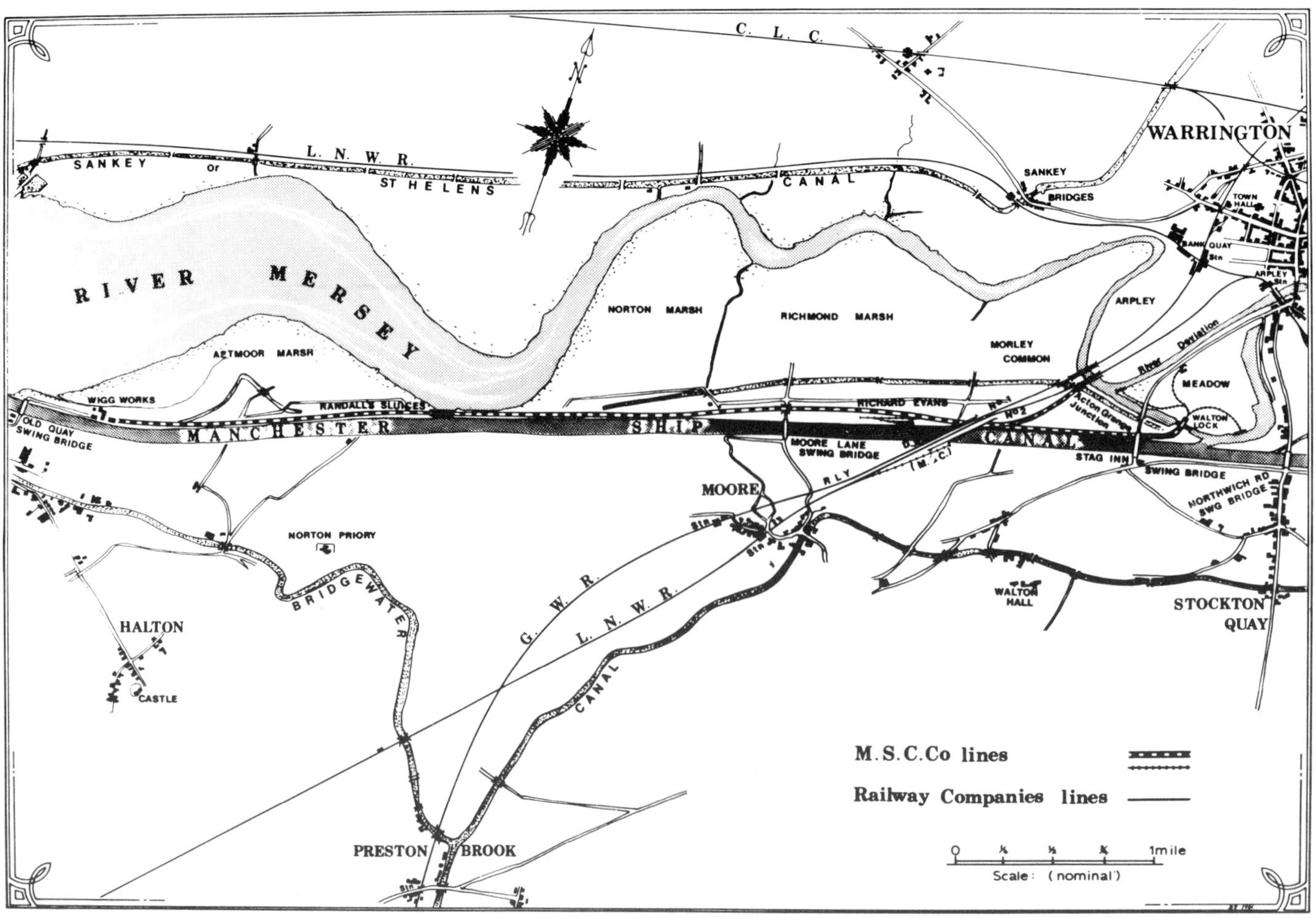
C. L. C.
N
WARRINGTON
SANKEY or ST HELENS CANAL
L. N. W. R.
SANKEY BRIDGES
TOWN HALL
BANK QUAY Stn
ARPLEY Stn
ARPLEY
RIVER MERSEY
NORTON MARSH
RICHMOND MARSH
MORLEY COMMON
ASTMOOR MARSH
WIGG WORKS
RANDALLS SLUICES
RICHARD EVANS
River Deviation
MEADOW
WALTON LOCK
OLD QUAY SWING BRIDGE
MANCHESTER SHIP CANAL
MOORE LANE SWING BRIDGE
STAG INN
SWING BRIDGE
NORTHWICH RD SWG BRIDGE
MOORE
RLY
NORTON PRIORY
WALTON HALL
STOCKTON QUAY
HALTON
CASTLE
BRIDGEWATER CANAL
G. W. R.
L. N. W. R.
PRESTON BROOK
M.S.C.Co lines
Railway Companies lines
0 ¼ ½ ¾ 1mile
Scale: (nominal)

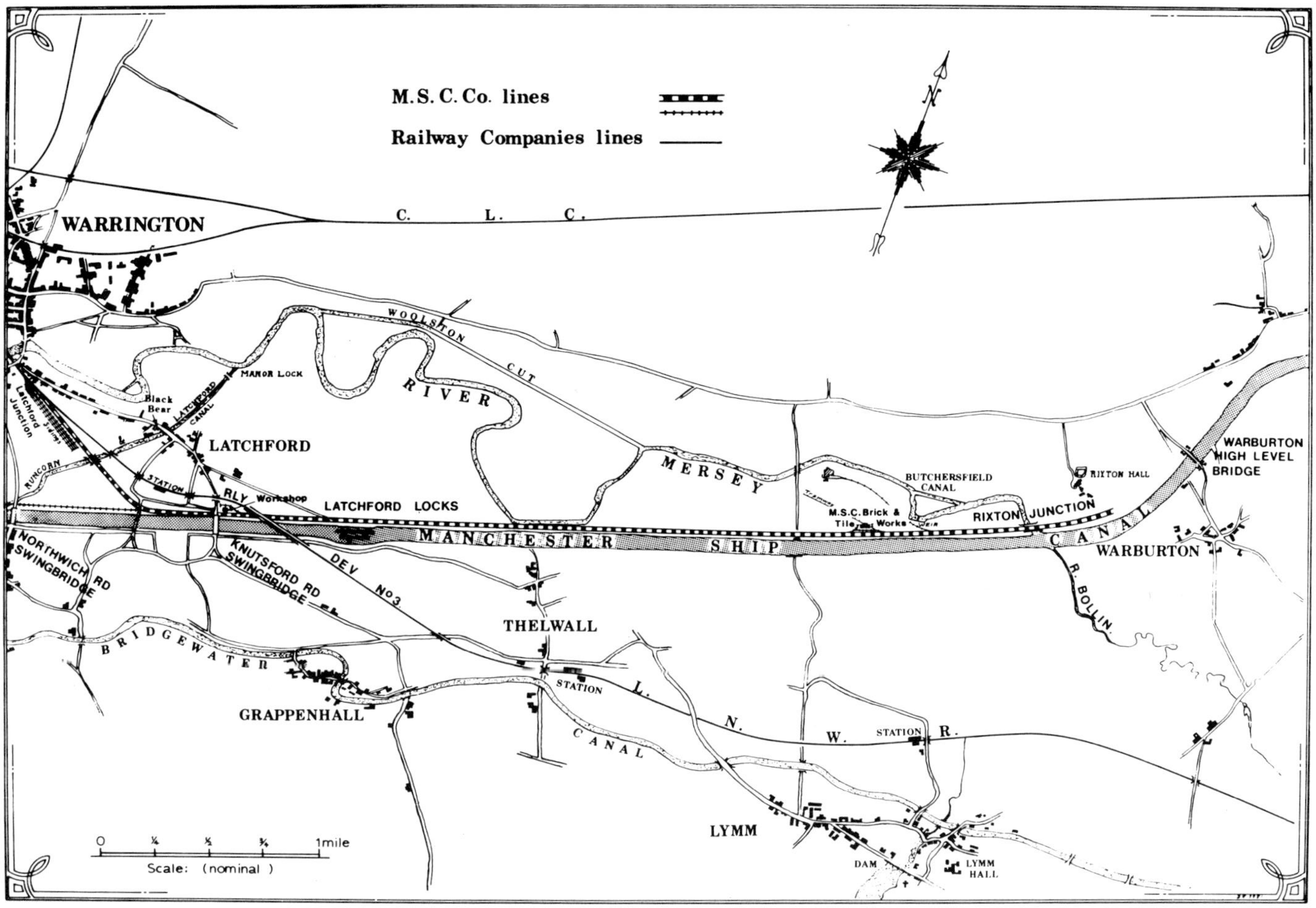
M.S.C.Co. lines
Railway Companies lines
N
WARRINGTON
C. L. C.
WOOLSTON CUT
RIVER MERSEY
MANOR LOCK
Black Bear
LATCHFORD CANAL
Latchford Junction
LATCHFORD
RUNCORN
STATION
RLY Workshop
LATCHFORD LOCKS
BUTCHERSFIELD CANAL
M.S.C. Brick & Tile Works
RIXTON HALL
RIXTON JUNCTION
WARBURTON HIGH LEVEL BRIDGE
MANCHESTER SHIP CANAL
WARBURTON
NORTHWICH RD SWINGBRIDGE
KNUTSFORD RD SWINGBRIDGE
DEV No3
R. BOLLIN
THELWALL
BRIDGEWATER CANAL
STATION
L. N. W. R.
GRAPPENHALL
STATION
LYMM
DAM
LYMM HALL
0 ¼ ½ ¾ 1mile
Scale: (nominal)

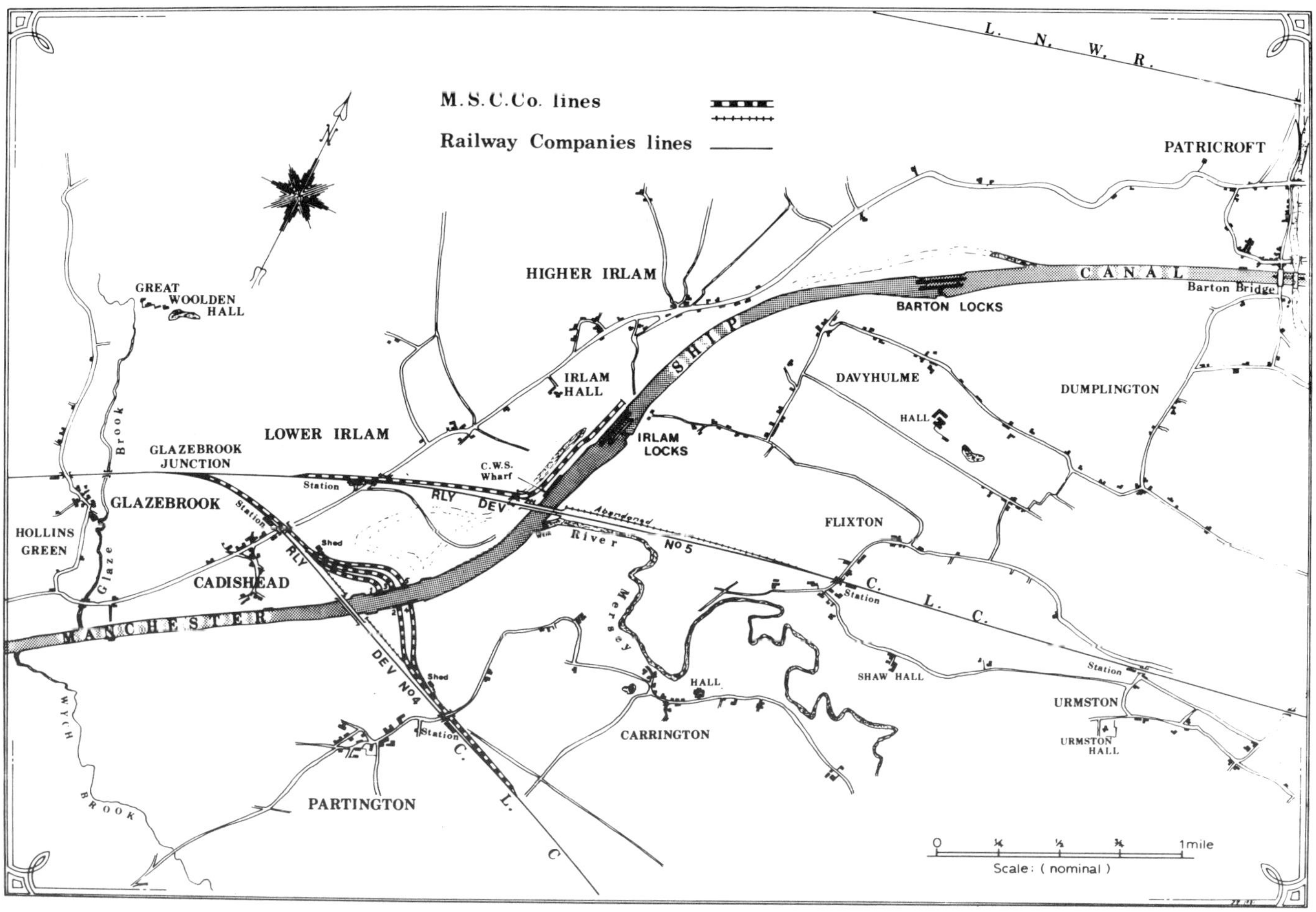
L. N. W. R.
M.S.C.Co. lines
Railway Companies lines
N
PATRICROFT
HIGHER IRLAM
GREAT WOOLDEN HALL
CANAL
Barton Bridge
BARTON LOCKS
SHIP
IRLAM HALL
DAVYHULME
DUMPLINGTON
HALL
Brook
GLAZEBROOK JUNCTION
LOWER IRLAM
IRLAM LOCKS
C.W.S. Wharf
GLAZEBROOK
Station
RLY DEV
HOLLINS GREEN
Station
Shed
RLY
River
No 5
FLIXTON
Glaze
CADISHEAD
C. L. C.
Station
MANCHESTER
Mersey
DEV No4
Shed
HALL
SHAW HALL
Station
URMSTON
URMSTON HALL
Station
CARRINGTON
PARTINGTON
BROOK
C. L. C.
0 ¼ ½ ¾ 1mile
Scale: (nominal)

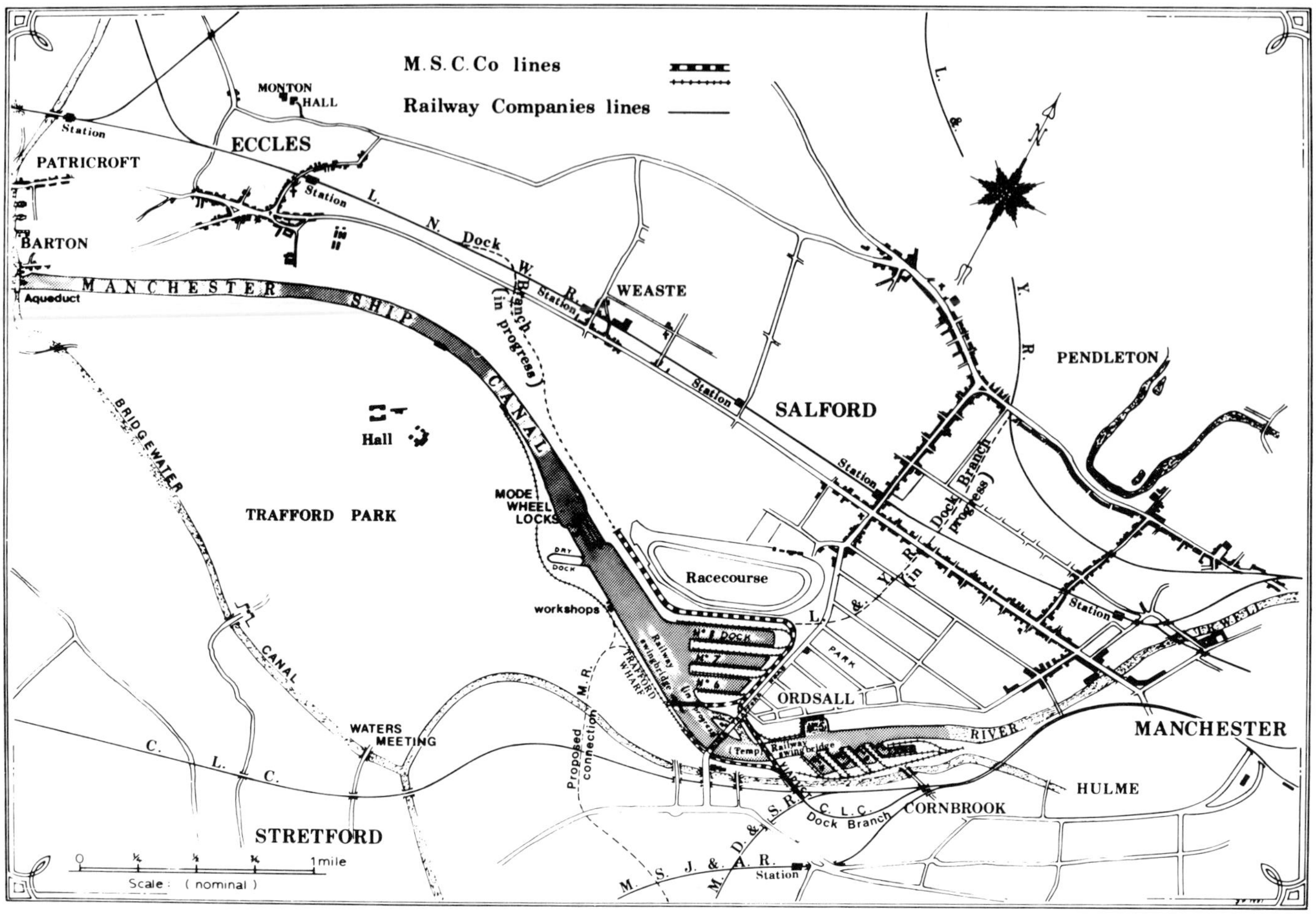
M.S.C.Co lines
Railway Companies lines
MONTON HALL
ECCLES
Station
PATRICROFT
BARTON
Aqueduct
MANCHESTER SHIP CANAL
L. N. W. R.
Dock Branch (in progress)
WEASTE
Station
SALFORD
PENDLETON
L. & Y. R.
Dock Branch (in progress)
BRIDGEWATER CANAL
Hall
TRAFFORD PARK
MODE WHEEL LOCKS
DRY DOCK
workshops
Racecourse
ORDSALL
PARK
MANCHESTER
RIVER
HULME
CORNBROOK
C. L. C. Dock Branch
WATERS MEETING
C. L. C.
Proposed connection
M. R.
TRAFFORD WHARF
Railway swingbridge
(Temp.) Railway swingbridge
M. S. J. & A. R.
M. D. & S. R.
Station
STRETFORD
Scale: (nominal)
1 mile

Chapter Two The Early Years (1894-1925)

Extent of the Track System 1894

As work on the construction had been nearing its end, most of the contractor's railway track had been taken up. What remained formed a system in eight isolated sections, each having a connection to the lines of the railway companies. The Company stated, at that time, that its railway connected to the main line companies lines at Cornbrook, Manchester (Cheshire Lines Committee), at Partington Coaling Basin North (CLC Glazebrook), and South (CLC Partington), at Latchford, Warrington (LNWR), at Acton Grange, Warrington (LNWR), at Runcorn (LNWR), at Saltport and at Ellesmere Port (LNWR/GWR Joint line), the latter two sections apparently sharing the same main line connection, one which lay on the east side of Ellesmere Port Station. It will be noted that only seven sections have been mentioned, the section not included by the Company being that which connected the line from Barton Lock to the new lines of the Cheshire Lines Committee (Deviation No. 5) using part of the lines of the ex-Cheshire Lines Committee, passing to the rear of the rebuilt Irlam Station to do so. Although this short stretch of line did not bear any industrial traffic at the time, and eventually was used solely by the CWS, it was part of the Manchester Ship Canal Railway until the section was closed seventy years later. The severed end of the lines of the ex-Cheshire Lines Committee, on the southern side of the Canal, was eventually lifted, the Company having had no plans for its use. This was also the case for the ex-LNWR (Deviation No. 3) line on the south side at Latchford, but the ex-LNWR (Deviation No. 2) line on the south side of the Canal at Acton Grange was retained for several years as a feed line for a proposed salt tip. This facility, which was to be sited near No. 1 and No. 2 Deviation bridge, never materialized and the connection was eventually taken up. Short stretches of contractor's line lay unused at Runcorn and Latchford, the latter stretch reaching from the old Latchford Station, by then taken over by the Manchester Ship Canal, to Northwich Road swing bridge. A short run of contractor's line which was still in use was that at Ordsall, Salford. It was laid around the incomplete No. 5 Dock and connected to the main Manchester Ship Canal Railway system in Salford Yard, via a line laid alongside a section of the River Irwell which passed under the original Trafford Road bridge. In 1894 that particular section of the old river course was being used as a bypass for small craft. The single track contractor's line was laid under the bridge on the northern side of the river course. Other stretches of track which could possibly have been originally part of the contractor's railway were the sections at Latchford, Acton Grange and that between Ellesmere Port and Saltport.

With the exception of the Eastham line, which was removed, the lines laid on the western side of Ellesmere Port Station were taken over by the Shropshire Union Railway & Canal Company to form its dock railway system. A small portion of an earlier tramway, from a point near Westminster Road, was incorporated after strengthening, and used as a siding.

Saltport

Details concerning this latter section, especially Saltport, are almost non-existent. In the previous chapter the birth of Saltport was described and a lack of knowledge expressed

Latchford Station on 28th May 1889. The nearest level crossing was that of Grappenhall Rd. (now Grammar School Rd., north side, Hilltop Rd., south side) whilst the distant crossing was that of Knutsford Rd. The latter position is now occupied by Knutsford Rd. swing bridge. Behind the group of figures is the station master's house, which is still in use as a residence, and to the right of the picture is the large station building which is also still standing but not in use. Four months after the completion of Railway Deviation No. 3, the lines and station became redundant and were purchased from the LNWR by the Manchester Ship Canal Co., the left-hand line afterwards being used as the MSC to Latchford Junction line.

Courtesy M S C Co.

The single line passing between the remaining buildings of the old Latchford Station. In the hazy distance is Knutsford Road swing bridge, whilst to the left of the picture is the embankment carrying the LNWR (Deviation No. 3) lines towards the viaduct spanning the Canal. The Canal can just be glimpsed to the right of the buildings. Compare this photograph with the one in Chapter 2 showing the station in full working condition.

G. Holt and A. Wadsworth

A much enlarged section of a photograph showing railway working at Saltport (date unknown). Of particular interest is the locomotive in front of a loaded wagon and alongside the warehouse. From the original photograph, the engine appears to be an ex-contractor's 0-4-0 machine with outside cylinders. An interesting feature of the locomotive is the rounded sides of the cab roof.

M S C Co.

A Smelting Corporation well tank locomotive, No. 4, and its crew pose at the Manchester Ship Canal wharf on Stanlow Meadows. Built by E. Borrows & Sons of St. Helens in 1898, the engine (Works No. 37) is preserved today on the North York Moors Railway. The locomotive driver is believed to be Mr A. Jones (senior) of Ellesmere Port.

Courtesy A. Jones

regarding its main line railway connections. However, the situation may be made a little clearer when the available facts are studied. The ground between the Canal at that point and the nearest main line, the LNWR/GWR Joint line, was marshy at first with more solid ground rising quite steeply, from a railway engineer's viewpoint, towards the main line at Frodsham. The nearest reasonable access point was near Helsby, but there is no record of any abandoned railway banking in any of these directions, either in early maps or visually. Finally, but significantly, there was no Act of Parliament granted to make such a connection in that area.

The truth is probably more mundane but, nevertheless, quite interesting in that the railway used would be the contractors' line which was laid alongside the Canal workings and which connected to the LNWR/GWR Joint lines at Ellesmere Port. Certainly the line running in the direction of Ellesmere Port branched off into the contractors' depot on Frodsham Marsh and it could be assumed that the locomotive or locomotives were stationed at the depot. By 1899 the railway at Saltport consisted of a 1½ miles length of track and sidings on the completely isolated embankment between the River Mersey and the Canal, and just over 1 mile of track on the Frodsham Marsh side, with sidings at Saltport and those into the ex-contractors' depot, formerly 'Marshville' by then known as Frodsham Marsh Huts. It would seem that the port was still being used but only as a transhipment facility for those ships, usually the tall-masted sailing ships, which could not proceed beyond Runcorn Bridge. Sailing ships were, by then, in their death throes, and this situation probably added to the early demise of Saltport in about 1901 or 1902.

Ellesmere Port

At Ellesmere Port the Company was operating a coal wharf with one tip in use. Cattle was also landed on occasions but this does not appear to have been a regular operation. Locomotive power was probably obtained from the Frodsham Depot. By 1897 Manchester Ship Canal interest in the whole section seems to have waned and by 1898 the coal wharf and its railway connection had fallen into disuse. Apart from the disused line to the wharf there was no line laid in the direction of Saltport. There is no date available for the closure of the Saltport-Ellesmere Port line but it probably occurred before the end of 1897. There is also no record of which locomotive or locomotives worked in the area after 1894, and a suggestion that the ex-contractors' engine *Grange* worked there cannot be verified but could, nevertheless, be correct, as that particular locomotive, an 0-4-0ST, built by Manning Wardle, was delivered to the contractor at Ince in 1887 and was employed in the Ellesmere Port-Saltport section.

A development occurred in 1898/9 which marked the rebirth of Manchester Ship Canal Railway activity at Ellesmere Port. This was the laying of a new line, ⅜ mile, into the site of the newly-established Smelting Corporation's Works, and the building of a wharf for the Smelting Corporation's own use. The line followed the same path as the earlier coal wharf line. No Manchester Ship Canal engines, rolling stock or staff were employed at Ellesmere Port, the railway being operated by the Smelting Corporation staff using their own locomotives.

However, by the end of 1901 the Smelting Corporation was in the hands of the Official Receiver, and in this state carried on until 1912 when the site was taken over by the Manchester Ship Canal Portland Cement Company, which had no connection with the Manchester Ship Canal. This company also used its own locomotives and staff. By that time the line had been extended, on the western side of the works, towards the Canal and terminated at a large wharf, and this appears to have been connected with the ex-Smelting Corporation system. Also during this period, a line had been laid alongside the LNWR/GWR Joint lines embankment, descending on to Stanlow Meadow, and after 1 mile in a easterly direction it veered towards the Canal, ending up in the dye making works of Messrs Meister, Lucius & Bruning. The new works was established in 1908 and the product was indigo dye. Again the firm moved its traffic over the Manchester Ship Canal lines employing its own locomotives and staff.

Manchester and Salford

Each of the remaining three sections of railway was purpose built. At Manchester and Salford the rails were set flush with the quays and cartroads in the usual way, with the feed lines, through lines and sidings laid on the surface. The system covered Pomona Docks, Trafford Wharf, Ordsall, Docks 6 to 8 and storage grounds alongside, and to the north of Mode Wheel Locks.

Partington

Partington Coaling Basin was completely railway-orientated. There were four coal tips in use, two on the south side and two on the north side. Also on the north side there was a steam crane for coaling small vessels. The sole purpose of the section was the handling of coal into ships. No facilities were available for the handling of imports.

Runcorn

The situation at Runcorn, in 1894, was very much as it was prior to the building of the Ship Canal but with the added facility of constant deep water quays. Cargoes handled were coal, cinders, china clay, salt, pitch pine, pottery and felspar.

It may come as a surprise to some people to learn that there was a docks railway system at Runcorn, but they would be even more surprised to learn that it had been in use for something like sixteen years before the Manchester Ship Canal was even started.

The development of Runcorn can be said to have started in 1772 when it became the site of the outlet of the Duke of Bridgewater's barge canal, allowing passage of wide barges between Manchester, the Duke's coal mines at Worsley and the River Mersey. In 1803, on the death of the Duke, the Bridgewater Canal and its docks at Manchester and Runcorn were put under the management of the Board of Trustees.

Apart from coal from the Worsley mines, Runcorn eventually was handling a great deal of traffic from the Midlands, and the Potteries in particular, the Bridgewater Canal by then being connected to the large network of narrower canals which had spread across England. In 1861, with the 'canal mania' dead and the 'railway mania' in full cry, an Act was granted to the LNWR for the building of a railway between the Warrington & Garston Railway (LNWR) at Widnes, and the Grand Junction Railway (LNWR) at Ditton, to pass through Runcorn and cross the River Mersey on a high level bridge. The contract for the railway and the bridge was awarded to Thomas Brassey and a start was made in 1863. On 1st February 1869, the railway across the Mersey was officially opened when a train hauled by the locomotive *Cheshire* travelled between Runcorn and Widnes.

With the arrival of the railway at Runcorn, the Bridgewater Trustees decided that a docks connection would be useful and

A view of the temporary railway swing bridge and the partly-constructed wooden gantry in 1893. These structures carried the CLC/MSC lines between Cornbrook and Ordsall in the early months of the Canal's life, the permanent railway swing bridge not being completed until the spring of 1895. In the background is a low-level wooden bridge which possibly could be the one which carried the contractor's railway over the River Irwell during the construction period.

M S C Co.

in March 1869 the Chief Engineer placed an order with Barninghams Iron Works, Pendleton, Salford, for lines and chairs. The actual date when the docks railway came into use is not definitely known but it could be circa 1870. It was certainly in use by 1872 when the Board of Trustees sold the Bridgewater Canal, the terminal docks at Manchester and Runcorn, also Duke's Dock Liverpool, to the Bridgewater Navigation Company, in reality GCR and MR people. Under this body the canal and docks remained until sold to the Manchester Ship Canal Company in 1887.

Little is known about the railway operation and the locomotives in pre-Manchester Ship Canal days, but one engine, ex-Dursley Railway No. 2020, an 0-4-0ST with 9in. x 16in. inside cylinders, was purchased, in December 1913, by the Bridgewater Navigation Company, from the MR for the sum of £400. Engine No. 2020 could have been the E. B. Wilson engine, works number 528 of 1856. The names of other locomotives reputed to have worked at Runcorn were *Wakefield*, *General* and *Earl*, though the latter two names could be one engine, *General Earl*. There was one such engine a Manning Wardle No. 965 of 1886. It was delivered new to Monk & Newall of Brindle Heath, Salford, and was an 0-4-0ST with 10in. x 16in. outside cylinders. When the docks at Runcorn were reopened for shipping on 9th June 1893, it is possible that the ex-contractor's engines *Latchford* and *Cheadle* were then providing the motive power, for they were both at Runcorn very early on in the Manchester Ship Canal era.

Railway Connections to Railway Companies Lines

The Manchester Ship Canal is referred to as an elongated port with jetties and docks at intervals along its length, but whatever is said about the benefits it brought to other towns, its main purpose was to bring trade to Manchester. This was reflected in the docks themselves. Manchester Docks, which means Manchester and Salford Docks, was paraded as an example of what a modern docks system should be. The beautifully executed masonry of the quays, the provision of a complete hydraulic system to the cranes and capstans of Nos. 6, 7 and 8 docks, and the large modern ferro-concrete warehouses, were all signs of a well thought out and engineered system. The Company also boasted of its railway system, yet on that opening day in 1894 the only connection between the docks railway at Manchester and the lines of the railway companies was through the contractor's temporary connection at Cornbrook, the permanent Cheshire Lines Committee connection not being ready until 4th January. Even then the situation was not a lot better, as the railway swing bridge carrying the Manchester Ship Canal line across the Canal from Trafford Wharf to Salford Yard was still in the course of construction, and therefore all railway traffic had to cross on the temporary single line swing bridge which connected Cornbrook with Ordsall Yard. The permanent swing bridge came into use during the late spring of 1895. Both temporary and permanent railway bridges were built by the Leeds firm, Messrs Butler & Company

Apart from the difficulties at Cornbrook the LNWR and L&YR were also behind in their connections. The LNWR was building its connection at Weaste and it was not completed

Railway swing bridge No. 1, which carried a single line between Salford yard and Trafford Wharf. Built by Messrs Butler & Co., Leeds, it was in use from early 1895 until 5th April 1943 when it was replaced by the present double line swing bridge.

Courtesy Trafford Park Estates Ltd.

until 4th November 1895. As was stated previously the LNWR was amongst the most bitter of the Canal's opponents, and its directors had delayed their attempts to obtain an Act authorizing a connection with the Manchester Ship Canal system until it became obvious that the Canal was really to be completed. In the case of the L&YR, its directors had seen the possibilities from the outset and had obtained an Act two years earlier than the LNWR, but with all the problems encountered in the construction of its entry into the docks, it was 28th March 1898, 2½ years after the LNWR had completed its connection at Weaste Junction, before the L&YR connection at New Barns was opened. Whereas the LNWR entered the Manchester Ship Canal system in a conventional fashion, through a cutting which passes under the LNWR, former L&M, lines and Eccles New Road, later the A57, the L&YR was forced to take its lines under the houses and streets of a densely-populated part of Salford. This entailed the building of several tunnels, bridges and cuttings over a distance of 1 mile 104 yards.

From Windsor Bridge Junction the lines were laid in the open to Liverpool Street after which they entered a 471 yard tunnel on a gradient of 1 in 47. This took the lines under the LNWR, former L&M, lines, West Egerton Street, and Regent Road before passing between the retaining walls of a cutting, after which they entered the 291 yard Ellesmere Street Tunnel. Beyond this was an open stretch spanned by 3 bridges, after which the lines plunged into the final tunnel, passing under West Park Street and Trafford Road before entering the docks close by Aubrey Street. The final tunnel was 172 yards long with a 1 in 68 gradient. Other streets affected were Martha Street, Tatton Street and Robert Hall Street and quite a number of buildings needed underpinning.

A rather interesting feature in the final tunnel was the passenger station which was built to handle patrons to the nearby Manchester Racecourse. Actually, the station was only used for this purpose until 1902 when the racecourse was purchased by the Manchester Ship Canal Company, and from that time, until 1939, was used very infrequently. The contracts for the L&YR connection and the station were awarded to Thomas Wrigley. On the docks the L&Y line joined the Manchester Ship Canal system at a gridiron, later No. 1 Grid, near Salford Quay. There was certainly a variation in the way railway connection were made at Manchester, underground to New Barns Junction, on the surface to Weaste Junction and, as previously described in Chapter 1, overhead to Cornbrook Junction.

Cornbrook Junction, which was believed to be single track over the viaduct, was rather unsatisfactory from the beginning and was the cause of some friction in the early days. During February 1894 traffic at Cornbrook came to a stop, the Cheshire Lines Committee refusing to deliver to the docks and the L&YR refusing to accept traffic from the docks. This happened soon after the Cheshire Lines Committee agreed to allow both L&YR and LNWR traffic to use the connection whilst their own connections were being constructed. The Manchester Ship Canal Company pointed out the fact that both companies Acts would not allow them to behave the way they were and, furthermore, that the whole affair would be put before the Railway Commissioners. One of the railway companies' spokesmen stated that the process was inconvenient but of little consequence and that the Manchester Ship Canal Company were being a bit awkward themselves. After that the matter was resolved and traffic passed through as before.

The viaduct was to have a relatively short life, for on 1st March 1905, the new Cheshire Lines Committee connection at Trafford Park Sidings, Stretford, came into use, Cornbrook Junction being closed at the same time. From Trafford Park

The first Lancashire & Yorkshire Railway's engine, and six-wheeled coach to enter New Barns Junction, Salford, in 1898. Although rather indistinct, the photograph shows the 0-6-0 tank engine dressed with bunting.

M S C Co.

An extra boost was given to the Port of Manchester when the new shipping line, Manchester Liners Ltd., began operating from the infant port. Pictured is the steamship *Manchester Enterprise* (formerly *Queensmore*) at Salford Docks. Unfortunately for the Railway Department, the early vessels did not bring much traffic as most of it, being local, was moved by horse-drawn carts. In fact, when the *Manchester City* docked on 16th January 1899 with 7,500 tons of mixed cargo, little of which was taken by rail, the Railway Manager was heard to remark, 'I hope you will bring something more for me next time'.

Courtesy Norman Edwards Associates Ltd. for Manchester Liners Ltd.

Sidings the line crossed over the Bridgewater Canal into the Trafford Park Estates railway system. The connection has, for that reason, been referred to as Bridgewater Junction. The incoming line reached the Manchester Ship Canal system at Trafford Wharf via a set of sidings later known as 'A' Junction. Although there was, and still is, a stiff gradient involved in the approach to Bridgewater Junction, it is nowhere near as bad as that on the old Cornbrook Viaduct.

Traffic

The volume of traffic passing through the docks in the opening months of 1894 was not great, and the railways share of it was very small compared with that which was moved by horse-drawn carts. In fact, it has been stated that only two trains each day were being moved at Manchester Docks, one inwards and one outwards. The number of wagons moved each week rarely exceeded 450. Railway tonnage for the whole of the Port of Manchester during 1894 amounted to only 411,921. It was to take 3 years before the businessmen of Manchester fully realized the value of the new Canal. There was also a matter of business and family ties between the Manchester and Liverpool interest, ties which added to the difficulties being experienced by the Company in attracting trade.

However, by 1900 the Port of Manchester was being used more fully and it must have been with a great sense of satisfaction that the Board of Directors surveyed the progress made. Shipping tonnage, carried by larger ships than was originally envisaged, had climbed steadily, and by then Manchester boasted its own shipping line, Manchester Liners, established 5th May 1898, the leading men behind the new line being Sir Christopher Furness and Mr R. P. Stoker, first Managing Director.

Expansion

In the autumn of 1899 a start had been made with a payment of £50,000, on repaying the interest on the Company's loans, the total amount overdue being £1,043,000.

However, the directors were confident of the future and were planning a major expansion of the docks and railways. The number of trains entering or leaving Manchester docks numbered about 30 each day. There were 590 of the Company's wagons entirely occupied around the docks whilst the railway companies were supplying 120,000 wagons. The docks railway totalled 45 miles and track mileage for the whole Canal was given as 95 miles. Staff had risen to 615 men and boys, occupied at 5 sections, or 'stations' as they were then called. The total railway tonnage for 1900 was alternatively given as 1,721,879 *(Port of Manchester 1708-1901)* and 1,682,182 *(Railway Magazine 1903).*

In the first year the only marshalling yard at Salford was on the southern side of 6 Dock (it was always referred to as 'South 6') and until 1901, when its capacity was doubled, it held a mere 250 wagons. The track being used for the whole of the Canal was given as 24 miles.

As the docks became busier and as more firms, shipping and manufacturing, established themselves on the Canal estate, the railway expanded. Most of this occurred at Salford, Eccles and Irlam. At Weaste the through route was laid during the 1894/5 period and this ran alongside the Canal from Mode Wheel Locks to Weaste Junction. From Weaste Junction the line was being laid towards Eccles, Irwell Park (later Eccles Oil Wharf) and to Barton Aqueduct. By 1898 the line was completed, but due to difficulties involved in putting it through Barton Bridge-Aqueduct area, it came to a dead end by the hydraulic tower. At Partington North the line had been extended eastwards by 1896 to connect with the line to Irlam Locks. It was laid alongside the abandoned course of the River Irwell, passing under the Cheshire Lines Committee (Deviation No. 5) lines and the lines of the ex-Cheshire Lines Committee with which it was to connect. At a later date, thought to be about 1912, the line was made redundant as the new through line, which was laid alongside the Canal, had come into use.

Railway expansion at Salford was authorized under the Act of 1896 which allowed them to 'make railways at Salford', and by the Act of 1900. Under the Act of 1896 a permanent line was laid from Salford Yard to Ordsall Dock (No. 5 Dock) and this followed the route of the ex-contractor's line under the old Trafford Road bridge. Included in the Act of 1900 was provision for the compulsory purchase of Manchester Racecourse which was to provide space for the construction of the proposed No. 9 and No. 10 docks. (As a point of interest No. 10 dock was never built). In 1899 a start was made on the construction of a large engineering works and locomotive shed at Mode Wheel. At Pomona Docks a railway bridge had been installed which spanned a corner of 2 Dock thus allowing the line to be put through to Cornbrook Wharf. Pomona Docks

Railway Deviation No. 5 Viaduct, carrying the CLC lines over the canal at Irlam. Originally the earth embankment carrying the lines up to the viaduct swept straight down to the edge of the Canal bank but, at the turn of the century, when it was decided to shift the Manchester Ship Canal through route from its position alongside the old River Irwell course to a position alongside the canal, a slice was taken out of the embankment and a retaining wall (seen on the left of the photograph) was built to hold back the soil.

D. Thorpe

A busy scene at Railway swing bridge No. 1, sometime between 1902 and 1914. The sidings on the opposite bank were, in the early days of the Canal, the only marshalling facility available. On the bridge is Hunslet engine No. 14 *St. John*, a 'Jazzer', whilst another 'Jazzer', No. 15 *New York*, can be seen on the left of the photograph. Behind the coal stage is an unidentified engine of the class, whilst on the right of the picture is a fourth 'Jazzer', believed to be No. 17 *New Orleans*. On the same line is a Hudswell Clarke engine, believed to be No. 39 *Sydney* and a Peckett, No. 12 *Jaffa*. *M S C Co.*

The single line which skirts the masonry under Barton Aqueduct. It will be seen that the line is at a very low level compared with the surface of the water. In the foreground is the casing containing the pumps used to pump the section dry after heavy rain. *D. Thorpe*

were very cramped with the result that only short chain curves were permissible which limited the size of locomotives able to work there. Also in the 1896/7 period the line at Partington North had been extended westwards to meet the single line to Latchford. Some time between 1903 and 1905 the line ending at Barton Locks was extended towards Barton Bridge, coming to a dead end in Barton Yard, an engineer's storage area situated alongside Peel Green Road, near the Rock Hotel. Less than a quarter of a mile separated the dead ends of Barton Yard and Barton hydraulic tower. Eventually, on 16th July 1913, the lines were connected after engineers had built a ledge out into the canal, taking the single line under the aqueduct and under a newly built bridge which carried the approach road to Barton Bridge. With the completion of this section the through route from Trafford Wharf to Latchford totalled 16½ miles. It is possible that the belated laying of the continuous line between Irlam and Salford was hastened by the growth of industry at Irlam. Apart from the CWS Soap & Candle Works, built in 1896, there was the Irlam Steel Works of the Partington Steel & Iron Company, construction of which began in 1910. The steel works was to have a tremendous impact on the Manchester Ship Canal Railway, both in tonnage moved and, of course, on staff and locomotive requirements.

During 1896, Manchester Corporation Cleansing Department commenced negotiations with the Company for the cutting through of the Canal bank to give access to part of the old River Irwell course at Boysnope Farm; an area between Barton Locks and Irlam Locks. A wharf was to be built on the old river course and a depot established. There were to be offices, stables, workshops and a locomotive shed at Boysnope Wharf, as it became, and from there emanated the Cleansing Department's narrow gauge, steam-operated railway system serving Chat Moss and Barton Moss. To cut through the Canal bank involved the construction of a single track bridge to carry the Company's railway estimated by the Company at £612 5s 10d and at £500 by the Corporation. A spur was to leave the Manchester Ship Canal line and enter Boysnope Wharf, initially to bring in building materials but eventually being used mainly for the delivery of coal.

As with many of the Company's negotiations with others at that time, other factors were thrown in with the object of gaining advantages for the Manchester Ship Canal. In this case it was the use of Corporation land at Barton for the dumping of dredging spoil.

Trafford Park and 9 Dock

Earlier important developments included the emergence, in 1896, of Trafford Park as an industrial estate, and the commencement of 9 Dock. Manchester Racecourse was closed on 1st November 1901 and the contractors moved in to construct what was, at 2,700ft. by 250ft., the largest dock at Salford. On 13th June 1905 it was officially opened by Their Royal Highnesses King Edward VII and Queen Alexandra.

Between the new dock and 8 Dock a large marshalling yard had been laid, No. 1 Grid, which was the previously mentioned No. 1 Grid, and was often referred to as New Barns. The area was divided into two grids, one for the Manchester Ship Canal's use, the other for the use of the L&YR. Originally 1 Grid was planned to accommodate 3,000 wagons but the resultant sets of siding would, in fact, only accommodate 1,450 wagons, the larger section being that of the Manchester Ship Canal which held 750 wagons. Down the southerly side of the siding lay the original L&YR incoming lines, and on the northern side stood the Manchester Ship Canal locomotive shed (New Shed).

The Through-route

With the purchase of the racecourse the way was then clear for a new double track through route, which was laid on the northerly boundary, running alongside the racecourse outer wall, and which thus avoided passing through the busy docks area. To pass over the incoming L&YR lines at New Barns it had been necessary to build a bridge which put a couple of 1 in 40 gradients into the Manchester Ship Canal lines, which were to cause a great deal of inconvenience for the Manchester Ship Canal for many years.

After leaving the old racecourse area the new lines were laid alongside Weaste Cemetery and Mode Wheel workshops to reach Weaste Junction before carrying on towards Partington. Incidently, Weaste Junction was always known by the local folk as 'the Deaf and Dumb Railway', the story being that, during its construction, such was the haste by the LNWR to connect to the Manchester Ship Canal system, the gangers on the job were told to sack men instantly if they were caught talking or listening to a conversation.

Developments at Latchford

During those years of continuous development at Salford, events elsewhere had been happening at a much slower pace. Railway traffic passing through Latchford Junction was confined to that of the Manchester Ship Canal Brick & Tile Works at Rixton, and the Statham Chemical Company, whose works was situated on the banks of the River Mersey about ½ a mile west of the aforementioned brickworks. However, in 1905, the Mostyn Tube Company established a factory near Latchford Locks for the production of copper-lined tubes and boiler mountings. Large quantities of strip steel and copper were brought into the works and the finished product was taken out, the Manchester Ship Canal railway handling the major part of the total tonnage, and fuel was also carried in large quantities. At an unknown date the closure of the Statham Chemical Company's works and the Manchester Ship Canal Brick & Tile Works took some traffic from the section, but there was still sufficient to make the railway viable. There was also the necessity of taking coal to Latchford Locks boiler house.

Acton Grange

The situation in the Acton Grange section was rather more complicated than that at Latchford. In the first place the railway system was split into two parts, the larger part being on a stretch of land from Warrington Wharf to Old Quay, Runcorn, which was bounded on the southerly side by the Canal, on the northerly side by the widening River Mersey, and to the east by the old River Mersey course, an area which is sometimes referred to as 'the island'. The railway system on the smaller part was reached by crossing from the 'island' over Walton Lock via a swing bridge which pivoted on the northerly wall of the lock itself. The swing bridge, which was completed in 1895, was turned to span the lock during the night when vessels would not be passing through between the River Mersey and Canal. The area of land was the southern segment of Arpley Meadow, cut through by a river deviation during the construction period, and had been designated as the site for the proposed Warrington Dock. On the section were

An aerial view of No. 1 Grid with No. 9 Dock to the left and No. 8 Dock on the right. The full grid shown is the Manchester Ship Canal section. To the left, at the top of the picture, is New Shed or No. 9 Dock Shed as it is sometimes referred to.

M S C Co.

New Shed or No. 9 Dock Shed on No. 1 Grid. This is a rare view as usually there was at least one engine in the shed.

Courtesy H. Hudson

A rather damp scene at Salford yard near Middle Gate, Trafford Road. All engines seen in this pre-1900 photograph are ex-contractor's machines. In the foreground and to the left is *Heywood*, Manning Wardle No. 1082 of 1888, later MSC No. 8. To the right is, possibly, *Glamorgan*, Hunslet No. 396 of 1886, whilst in the near centre is *Withington* a Walker Bros. engine built in 1888. The saddle tank engine at the back is unidentifiable.
Stretford Library

offices and a locomotive shed, these being situated at a point near the new riverside road, later Chester Road. Facing the opening to the Canal was a wharf and the works of the Northern Creosoting Company, where railway sleepers and telegraph poles were treated. A line led to the wharf and a line was also laid which passed under the Chester Road (fixed) bridge and connected to a wharf on the River Mersey bank. It had been stated that a contractor's locomotive shed was sited close by this wharf and that it existed until 1910 or thereabouts.

On the 'island' was the chemical works of the Warrington Borate Company, which was sited near Walton Lock, the colliery wharf of Richard Evans, and the chemical works of Wigg Bros. & Steel at Old Quay, Runcorn. In February 1897 a contract was given out for the construction of a wharf and railway lines at the entrance to the proposed Warrington Dock. This was, in fact, the only docking facility ever gained by Warrington and later, after the town council had regretfully allowed the Manchester Ship Canal Company to withdraw the Warrington Dock clause from the Act, it became known as Warrington Wharf. With the abandonment of the dock plan the railway operation was moved to the 'island', and Arpley Meadow was given over to the Dredging Department for use as a spoil ground. The exact date this happened is obscure but it would appear to be circa 1909.

A new set of offices was built and later, in 1912, with the closure of the Warrington Borate Company's works, a new locomotive shed was erected, it is said from timbers taken from the abandoned chemical works. It is rather interesting to note that the Acton Grange engine has always been referred to, since about 1909, as the 'Arpley Meadow engine'.

At a later date a small sand quarry was established near the LNWR/GWR bridge (Deviation Nos. 1 and 2) and the Manchester Ship Canal put in and removed one wagon each day. A more profitable traffic was obtained with the arrival, in 1920, of the older established timber and woodworking firm of R. A. Naylor Limited. Their premises stretched, and still do, from Chester Road almost to Warrington Wharf. With the building of their new premises on the western side of Chester Road some years later, the two works became known as East and West works.

Expansion at Partington

During the same period there had been developments at Irlam and at Partington Coaling Basin. The Coaling Basin originally had about 7 miles of track, but as the section became busier, extra wagon hoists (coal tips) were installed in 1907 and extra sidings laid. By the turn of the century the track mileage had doubled, by 1906 it had reached 19½ miles and by 1920 the track layout had more or less reached its final form with a total of 24 miles. A certain amount of the increased track length was due to the establishing of Irlam Steelworks.

In the early months of 1911, additional sidings were laid solely for the steelworks use, whilst in 1912 the junction at the western end of the works was being made. Actually the junction should have been completed early in 1912 but a series of delays ensured that it was late summer before the work was completed. During the winter of 1913/14 further additions were made to the sidings and in 1915 additional sidings were laid at Glazebrook Junction.

Expansion at Irlam

There were some industrial developments at Irlam and Cadishead which did not show any significant increase to the track length but, nevertheless, added considerably to the

SS Mabel, a coasting vessel, taking on a cargo of coal from No. 2 Tip at Partington Coaling Basin. Note the wagon on the lower level about to be tipped, and the two wagons on the upper level ready to move off the tip.

Stretford Library

Walton Lock, Warrington, circa 1910, with a rather dilapidated railway swing bridge spanning the lock, and a fixed bridge spanning the sluiceway (foreground). On the far side of the lock is Arpley Meadow and just visible is one 'owners' van, therefore the section, most probably, was still in use.

M S C Co.

railway tonnage figures. The earliest manufacturing firm to establish its premises on the Canal estate, at Cadishead, was the Patent Fuel Company. Fuel briquettes were produced by the firm until it ceased trading in about 1918. In July 1913 the erection of a chemical factory was started at Irlam by a local building firm, Brew Bros. The ill-fated plant was being built for the Magadi Soda Company, the intention being to process soda, which was to come from Lake Magadi in East Africa, into calcium nitrate. Trouble commenced for the firm with the start of World War I in 1914, when, due to the German activities in East Africa, the raw materials could not be brought out of the country. Eventually, the Government took over the unfinished Irlam plant, but instead of allowing it to be completed, the Ministry for Munitions stopped any further work on the project. The plant was taken over on 17th February 1916 but was immediately dismantled and rebuilt elsewhere. From that time onwards the Manchester Ship Canal men have remembered the name of the unfortunate company, not because of the excellence of their memories, which many of them do have, but by a loop line put in to serve the plant and which was retained by the Manchester Ship Canal as a passing loop in the single line system. It was always known as the Magadi Loop.

About the same time that the sorry tale of the Magadi Soda Company was being enacted, Brew Bros. were also building a factory near the Patent Fuel Works, Cadishead. This was the Manna Oil Works and its produce was cattle feed cake. Unlike the soda works, the oil works was completed and went into production, lasting until circa 1935 when it was closed down.

Another CWS factory, the margarine works, was built in Irlam, the year being 1916. It was, and still is, situated about ¼ mile from the Eccles-Irlam boundary. A major expansion of the works occurred in 1920. Also in 1916 another firm established its premises at Cadishead. The name of the company was British Tar Products, and apart from processing tar the firm also refined benzole. A similar plant was set up about six years later by North West Co-operative Tar Distillers, later Lancashire Tar Distillers, its premises being sited further west of British Tar Products and close by the Cadishead boundary. Both tar companies and both CWS works installed their own railway system, using their own locomotives for internal shunting. Manna Works, however, used horses for internal shunting.

In January 1913 the opening moves were made by the Earl of Ellesmere for the extending of his colliery railway from the canal basin at Boothstown, to a point near Salteye lay-by, Barton, where the line would join the Manchester Ship Canal Railway. The original plan envisaged a set of sidings and two junctions, one near Barton Locks the other at Salteye lay-by.

This proposal did not suit Manchester Corporation, part of its land on Barton Moss and Chat Moss being sought after by the Earl of Ellesmere, the reason being the amount of Canal frontage being taken up. Plans were modified and a second set of proposals submitted on 13th April 1913, Plan 'A' having the line join the Manchester Ship Canal line at Salteye lay-by, and Plan 'B' showing the junction being formed in a westerly direction alongside Barton Locks. The urgency of the scheme was stressed several times. After five years polite negotiation, the matter was finally settled when, on 1st January 1918, a sum was deposited with Manchester Corporation as first part of the £13,787 15s 0d purchase price. The land became the property of the Earl of Ellesmere on 1st June 1918 when the balance, plus £65 6s 10d interest, was paid. A lot of time and a lot of money; and yet six years later, on 31st March 1924, only the foundations for the abutments and wings of the bridge planned to span Liverpool Road near Salteye, were completed. The scheme lay dormant until 1944 when it was resubmitted, but with the added facility of a junction to the LMS lines at Barton Moss. This attempt also came to nothing.

Industrial Growth at Manchester and Salford

Although the busiest section of the Canal was obviously Manchester Docks, there were relatively few factories sited on the company's estate in that area. Many of the new firms to the area were, of course, settling in the new and bustling industrial estate of Trafford Park. One section of Trafford Park actually breasted on to the Manchester Ship Canal Company's Trafford Wharf, and most of the premises along the wharf had sidings connected directly with the Manchester Ship Canal Railway system. Morrel Mills, ship repairers, Illingworth Ingham, timber importers, Hall & Pickles, steel stockists, these firms and the flour mills of the CWS and Hovis Bread Flour Company were amongst the earliest concerns to establish their premises along the wharf. The earliest firm on the Manchester Ship Canal estate at Trafford Wharf was the Manchester Dry Dock Company (1894) whose first pontoon facility had been set up at Ellesmere Port in 1893. Beyond the graving dock of the Manchester Dry Dock Company were the cattle lairages of Manchester Corporation. Most of the firms which had established their offices in the docks area itself, were, of course, shippers or shippers' agents with a sprinkling of warehousemen. At Pomona Docks this was also the case, the only exceptions being Arthur Guinness Son & Company, the brewers, who had established a depot and wharf at No. 1 dock in 1897 and brought the beer to Manchester in their own ships, A. Brown & Company, granite merchants, and The China Clay Company. Alongside the Canal, close by the entrance of No. 4 Dock, was the coal wharf of The Bridgewater Trustees.

However, on the opposite side of the Canal at Ordsall the situation was completely different. At least two factories were established there before the Canal had even been discussed. A firm of soap makers, Goodwin Bros. had built its premises, Ivy Soapworks, in 1877 whilst the Ordsall Hall Works of Chadwick & Taylor was built about the same time on a site near Trafford Road bridge. There was also a cotton mill owned by Richard Howarth & Company but the mill, in fact, was never connected to the Manchester Ship Canal Railway. New industrial concerns at Ordsall were Joseph Pyke & Son, Trafford Mill (1912) and John Jackson, Dock Mill, (1915) both corn millers. The latter mill was built on the reclaimed ground of the filled-in No. 5 Dock. In 1904 a firm of warehousemen, Distributor Limited., had built its premises near Ordsall Lane and alongside the soapworks. By 1909 the warehouse had been taken by another warehouse company, W. W. Jones Dooly. The year of 1918 saw the building of the tea blending and packaging factory of the English & Scottish Joint Co-operative Wholesale Society.

About half a mile away to the north of No. 9 Dock, at the corner of Montford Street and Vere Street, was the sawmill of William Evans & Company. Established about 1912 it was later purchased, along with adjoining land, by the CWS and became their Langworthy Saw Mill which also housed a large joinery section. Nearby, at the corner of Kellet Street and Langworthy Road South, in 1922, a packing warehouse and sawmills (New Barns Works) were built for Lloyds Packing

A train of insulated meat vans is seen, circa 1898, leaving the lairages of Manchester Corporation at Trafford Wharf. The photograph is particularly interesting in that it is one of the rare pictures of ex-contractor's locomotives (except Nos. 1 to 10) working for the Traffic Department. Built in 1887 by Manning Wardle, (Works No. 1027), *Grange* worked in the Ince to Ellesmere Port section during the construction period.

M S C Co.

The Anglo-American oil train, after leaving the depot on Trafford Wharf. Heading the train is Peckett No. 654, of 1897, *Alexandria*. It is worth noting that the depot was later taken over by Esso in 1954, this company also becoming the owner of the locomotive, by then MSC No. 11.

M S C Co.

Limited. Both the warehouse and the joinery works were served by rail, the lines passing over ground previously occupied by New Barns Farm which, incidently, gave its name to New Barns Junction which, with 9 Dock and the racecourse was also built in part of the farm. By 1899 the cold storage plant of The Colonial Consignment & Distribution Co. was being constructed at Mode Wheel and a special rail link formed. Beyond the newly built Mode Wheel workshops, a lime, sand, and coal merchant, James Roscoe (Bolton) Limited had, in 1900, set up a quarry. It was situated between Mode Wheel workshops and the newly laid Daniel Adamson Road. A line was put into the area, running three quarters of the way up Daniel Adamson Road. The firm finished operations circa 1913 but the sand quarry and the feed line remained much as it was for another thirty years. The Manchester Ship Canal sand quarry lay about half a mile west and on the opposite side of Weaste Junction in an area known as Little Bolton. Between the quarry and Weaste Junction were the tanks and sidings of the General Petroleum Company. Another oil works was built at Eccles for the Anglo Caucasian Oil Company. The works was completed and the first tanker unloaded in 1898. By 1912 the works was owned by the Consolidated Petroleum Company Limited.

Increase in Sidings

Apart from the growth of sidings into the various firms premises at Manchester, there had also been an increase in the Company's sidings. North of 9 Dock, a set of sidings had been laid, in 1906, providing accommodation for 220 wagons, whilst in 1909 a section of Irwell Park, which was being used for timber storage, also gained a new area of sidings. Two years later extra sidings were put in at Weaste Junction which increased its capacity to about 550 wagons, with further accommodation alongside Salford Corporation's Wharf. The LNWR sidings at Weaste held about 600 wagons with extra accommodation away from the Canal near Eccles Station. Over the years there were several minor extensions to the Manchester Ship Canal system, an important year being 1914 when fairly extensive additions were made to the sidings at Manchester, Irlam and Partington.

Development of Stanlow and Ellesmere Port

In 1916 the bridge carrying the line from Ellesmere Port Station over the Shropshire Union Canal was replaced by a new steel bridge. This was the first reported incident giving indication of the Company's plans for the development of Stanlow Meadow. Three years later a contract was awarded to the civil engineering firm, Sir Robert McAlpine to build a railway and docking facility at Stanlow. It was to be one of the most important developments for the Manchester Ship Canal Railway since the creation of the Trafford Park industrial estate. However, unlike Trafford Park with its mixed industries, Stanlow was to become the site of firms involved mainly with oil and chemical products. The estate eventually occupied land on both sides of the LNWR/GWR lines, and connecting both sections was a single line which shared an arched bridge with the aptly named Bridges Road.

Access to the Stanlow complex by rail was effected by descending a 1 in 76 gradient, 'the bank', on a single track which emanated from the Manchester Ship Canal's main line exchange sidings at the eastern end of Ellesmere Port Station. As there was already an existing junction at the western end of the station connecting the 'Shroppie' lines to the main line, the junctions became known at Ellesmere Port East and

A pre-1914 quay scene at Salford Docks with the *S S Locksley Hall* of Glasgow being unloaded, using hydraulic cranes. At the head of the train on the quay is an unidentified short tank Hudswell locomotive.

M S C Co.

Hudswell Clarke locomotive No. 644 *Belfast* (No. 28) at Salford pre-1914 with Driver Andrew Agnew near the centre of the engine. The fireman's name is unknown. Note the immaculate condition of the engine, including the highly-polished dome and manhole casings.

Mrs Hardy

One could be excused for mistaking this photograph for that of an Australian stock terminal with the open space, brilliant sunshine and the polished Hunslet engine. In fact the 'Jazzer' is at the head of a cattle train on Trafford Wharf on the Manchester Ship Canal. The date is unknown, but it could be around 1910.

M S C Co.

In Manchester Ship Canal railway folklore, *Hamburg* is quoted as being the locomotive responsible for the changing of the engine names for numbers. The story tells of patriotic dockers stoning the engine with its German city name (World War I was being fought at the time) and threatening to throw the luckless driver in the Canal if he persisted in bringing the engine on the docks. Incidents of this nature probably did occur, but the changing of names to numbers was actually for clerical expediency. Standing by the engine in the photograph is Driver Peter Agnew, one of the second generation Manchester Ship Canal Agnews. No. 31, as the engine became, is preserved at the Keighley & Worth Valley Railway, with nameplates restored.

Courtesy Mrs C Hignet

The railway bridge at Waters Meeting, Stretford, being floated into position during the early months of 1925. With the completion of this bridge, together with the bridge over Moss Rd., the line from the CLC Trafford Park sidings to Barton Dock Estate was able to be laid. Although of double track width, the bridge only ever carried a single track. The steelwork contract was awarded to Messrs Butler & Co., Leeds.

Courtesy W. Allen

Ellesmere Port West. Actually, although it is being referred to as the 'Shroppie', the dock lines of the SUR&C along with its docks were taken over by the Manchester Ship Canal on lease soon after the opening of Stanlow in 1922. As a point of interest, local people still refer to those lines as the 'Shroppie'. Near the station was a locomotive shed which originally housed the dock branch engines of the LNWR.

A road was built which stretched the full length of the new estate, and was aptly named Oil Sites Road. The Manchester Ship Canal lines were laid alongside Oil Sites Road which was crossed at intervals by feed lines to the various firms. Amongst these firms was the previously mentioned indigo dye manufacturers, at that time Levinstein & Company, later British Dyestuffs Corporation (I. C. I.). Between the through line and the LNWR/GWR embankment a large area of sidings were laid. Within a few hundred yards of Bridges Road and 'the bank' a shed was built and later a second shed was attached to it. Eventually both sheds shared a single roof. Cabins were also erected for shunters, number takers, etc.

On 1st April 1922 the Manchester Ship Canal started operating the docks at Ellesmere Port and within three months, on 22nd June, the junction at Stanlow opened. The situation as regards a split railway operation was similar to that which existed during the period when the Canal was being constructed. Movement of Manchester Ship Canal traffic, usually exchange locomotives, could only be effected by running through Ellesmere Port Station. As can be imagined, this situation raised some problems. Only those Manchester Ship Canal locomotives and drivers which were passed by the railway companies' inspectors were allowed to pass over the lines of the LMS, as the regional company became. Locomotives were inspected for clearances, tyre profile and brakes, the tyre profile being originally to GWR standard.

With the acquisition of Ellesmere Port Docks and its railway system, the Company also gathered four additional firms to be served by rail. Three were flour millers, their premises being sited alongside the dock basin. They were Imperial Flour Mills (est. 1903), King Flour Mills, (1905) and F. A. Frost & Sons, (1906). The fourth firm, Gabriel Wade & English Limited, were timber preservers, and the items treated with creosote were telegraph poles, and railway sleepers etc. The track length in both sections amounted to about 24 miles.

Staff was recruited locally with the exception of locomotive drivers who were brought in from Manchester. The first drivers to arrive at Ellesmere Port were Peter and Andrew Agnew, uncle and nephew. Locomotives were also transferred from Manchester, locomotive 71 being delivered to Manchester in December 1921 then taken by LNWR slow goods to Ellesmere Port on 10th July 1922. There was at least one other engine sent there, either before or after that date, but there is no record as to which locomotive it was. A further two engines, locomotives 66 and 70 were transferred to the new section on 1st December the same year.

The Beginning of Barton Dock Estate

At the same time that Stanlow was undergoing its transformation, the Company was making plans for another industrial estate at Urmston near Stretford. In 1919 Trafford Park Estates Limited began negotiations with the de Trafford Trustees for the purchase of land which lay on the western side of the Bridgewater Canal at Barton. However, the following year, part of that land was purchased by the Manchester Ship Canal Company and formed the basis of the Company's Barton Dock Estate. Between 1923 and 1929, Trafford Park Estates Company and the Manchester Ship Canal Company did a bit of 'land swapping', with the Manchester Ship Canal Company acquiring the former Lostock Farm and other isolated sections, in exchange for land on the Trafford Park Extensions at Barton. Another section, comprising some 37 acres at Dumplington, was sold jointly by Trafford Park Estates Company and Taylor Brothers. With the acquisition of this land the Company's Barton Dock Estate then covered an area of 1,000 acres.

The first firm to move into the estate was Messrs D. Anderson & Son Limited, manufacturers of roofing materials, its premises lying alongside the Bridgewater Canal and at the southern end of the estate. The plant was in production by 1921. An access road was laid from Park Road, Stretford but it was not until autumn 1925 that the Manchester Ship Canal Railway lines were put in. To enter Barton Dock Estate by rail, entailed a run through Trafford Park to Bridgewater Junction, passing through the Cheshire Lines Committee Trafford Park sidings and back over the Bridgewater Canal and crossing the new bridge near Waters Meeting at Stretford. The service line, which was single track, crossed over Park Road, splitting to form a line both sides of the service road, later Barton Dock Road.

Progress from 1894-1914

Looking back on the progress made by the railway, in the years up to 1914, is a rather inconclusive exercise because of the lack of figures regarding traffic moved between the years 1902 and 1913. There are, of course, the shipping tonnage figures for the same period and it can be assumed that the railways moved approximately the same percentage of the total each year. Demands made on the Railway Department were reflected in the increase in rolling stock and locomotives. Of the original 22 locomotives retained by the Locomotive Department from the contractor's fleet, about 10 remained in 1914, but they had been supplemented by a further 50 heavier locomotives, most of them purchased new from the makers. There were also 2,069 wagons owned by the Company, although most were purchased second-hand.

World War I

Trade had developed quite rapidly once the full potential of the Canal had been realized, but with the coming of World War I the shipping tonnage started to fall. However, for the railway, war created additional traffic with the tonnage rising from the 1913 total of 3,779,757 to a peak total of 5,593,635 in 1917. This increase was due to the particular function of the Manchester Ship Canal Railway as a service to the numerous industrial premises which, at that time, were turning out essential war products. Much of the traffic simply never touched the docks, it being moved between factories and the main line junction only.

The war came to an end on 11th November 1918 in victory for Britain and her allies. There were high hopes that peace would bring prosperity and, at first, that is how it appeared. Within five years however, the brief post-war boom was flagging, with the result that many firms were closing down and throwing thousands out of work. These were the first signs of the serious economic depression which was to grip the world for over a decade.

For the Manchester Ship Canal Railway the end of the war brought a sharp drop in tonnage as factories went through a change-over period from wartime to peacetime products. In 1922, as trade showed a recovery, the tonnage figures leaped to a peak which topped the peak wartime total, but then dropped as the first signs of the trade recession began to show.

Below The Railway bridge spanning the Bridgewater Canal at Waters Meeting. The canal can just be seen in the lower left of the picture. In the near distance, through the bridge, are Trafford Park sidings.
D. Thorpe

During 1925, the railway bridge at Moss Rd. was erected to carry the single line from the CLC Trafford Park sidings into Barton Dock Estate. Originally designed as a double track bridge, the brickwork was built as drawn but the steelwork contract, No. 227 awarded to Messrs Butler & Co., Leeds, specified north and central plate girders and north troughing only. The second half of the bridge was to be installed if warranted by increased traffic, but this did not happen and the bridge remains in its original form today.
D. Thorpe

Chapter Three
The Years of Consolidation and Strife (1925-1945)

Single Line Control System

On completion of the line into Barton Dock Estate, the railway had almost reached its peak with a route length of 33 miles and a track length of about 225 miles. The major part of the route mileage was laid single track, the longest stretches being between Barton and Latchford. The line between Barton and Partington was a particularly busy section and during a ten year period was remarkable in having more collisions and accidents than any of the other sections.

After one collision, in 1924, instructions were issued by Mr A. Alcock, the Railway Superintendent, in August of that year, to the effect that no engines were to be allowed in the single line section without first having received permission to do so. Even so, control of the movement of traffic was very lax and trains would sometimes pass through the sections close behind other trains. On 25th June 1925, locomotive 72, a Kitson, was taking a train of empties up a gradient near Barton Locks, whilst a short distance behind, at a suitable distance, ambled locomotive 29 *Bombay* travelling light engine. The usual method of ascending the gradient, Barton Bank, was for the engine to rush it in order to reach the top without stalling. For some reason locomotive 72 snatched at the train with the result that a wagon coupling parted and the train ran back downhill. Locomotive 29 was reversed in order to try matching speeds with the runaways and to attempt to bring them to a stop, but unfortunately events happened too quickly for such a manoeuvre to succeed, and the wagons hit locomotive 29 driving it back, but not before depositing two wagons on top of the engine! Incredibly no one was injured.

A further accident occurred, in February 1931, when a Kitson taking a coal train from Partington to Weaste, during foggy weather, collided with locomotive 73, another Kitson, which was travelling light engine. Both engines were travelling cab foremost and locomotive 73, on impact, was knocked back a considerable distance. Again no one was injured. Locomotive 73 was, in fact, on its way to Mode Wheel for overhaul when the accident occurred. Mr Alcock issued fresh instructions, on 20th February, regarding single line working but these merely reiterated previous instructions and added little to the safety of the system.

The earlier method of gaining entry into the single line system, from the Irwell Park section, was for the shunter to phone the Bridgemaster at Barton Bridge who had telephonic communication with Control (Coal Superintendents Office) at Partington. The Bridgemaster also had, on good days, a clear view of the railway line on both sides of the aqueduct and bridge. If no trains were in the section travelling towards Irwell Park from Partington, permission to proceed would be indicated by the signal situated at the start of the single line. This haphazard form of control continued until 1934 when a fatal accident forced the Company into adopting a better system.

The accident occurred at 08.30a.m. on Thursday 15th February 1934, on a day of thick fog. Due to the fog, the Manchester Ship Canal launch *Dispatch*, which was used by the cashiers when paying out along the Canal, was stopped from running. As was the usual practice on such days, the cashier and his assistant were taken along the Canal by engine as far as Latchford, where they transferred to hire car for the rest of the duty. A second engine would leave Salford an hour later carrying a second cashier and assistant, their duty only extending as far as Partington. It was this, the second cashiers engine, locomotive 56, which collided with a Kitson, locomotive 74. The Kitson was proceeding from Partington towards Manchester with a train of 5 loads and 16 empties, and the driver had asked the lock men at Barton Locks to telephone Barton Bridge informing the bridgemaster that they were coming through, and at the same time requesting that the second cashiers engine be held back. Unfortunately locomotive 56 had already left Barton Bridge and was proceeding towards Barton Locks and the oncoming train. Apart from the footplate staff and the cashiers, there was a sixth person on the engine, and he was standing outside the cab in front of the tank. This man, a rigger on his way to a job at Barton Locks, had begged a lift from the driver who allowed him to climb aboard, although it was against the regulations to do so.

Locomotive No. 81, Kitson & Co. No. 5422, built in 1929, is pictured at Ellesmere Port, possibly, by its condition, when new.
Courtesy Mrs C Hignett

Locomotive No. 29 *Bombay* with an unauthorized load. This rather spectacular balancing act was the result of the collision described in the text, when the engine was hit by a breakaway train at Barton Hump in 1925. The chimney can just be seen rammed up against the dome. Luckily, nobody was injured in the accident. MSC wagon No. 1214 was rebuilt to be used until scrapped in 1950.

Courtesy H. North

Locomotive No. 34 with a brake van at Moore on 19th May 1928. This engine, built in 1880 by Hudswell Clarke & Rodgers, (Works No. 218) originally started life as No. 6 for J. J. Charlesworth, colliery owners, and on being purchased by the Manchester Ship Canal in 1904, was named *Copenhagen*.

Courtesy A Appleton

One of the few private firms to have running rights on the Manchester Ship Canal Railway was the Partington Steel & Iron Co. Ltd. Running rights existed for the stretch of line between Irlam Steelworks and the Company's slag tips at Hollins Green and Rixton. No. 13, a Peckett engine (Works No. 1463), was built in 1917. The photograph is believed to have been taken soon after the locomotive was purchased.

Courtesy Urmston Model Engineers' Society

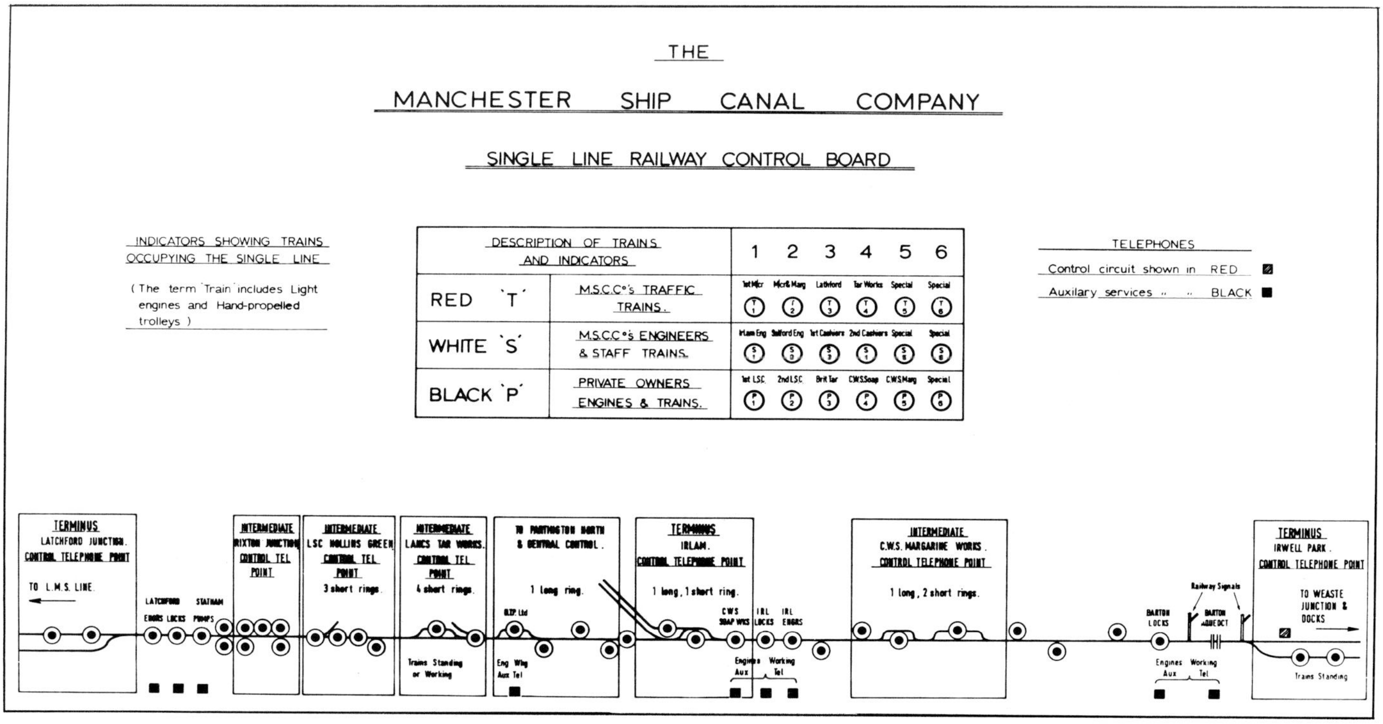

Visibility was down to 10 or 15 yards and the engine was moving slowly forward, all lamps lit, when out of the fog loomed locomotive 74. The impact was so violent that locomotive 56 was knocked back about 25 yards, and landed with all six wheels off the track and about 14in. away from the rails. The luckless rigger was flung against the tank, striking it with his head and he was taken to hospital, where, nineteen days later, he died of a brain haemorrhage. With the exception of one of the cashiers, the injured men were taken to Eccles and Patricroft Hospital. Of the superficially injured, only the shunter of locomotive 74, who had broken both wrists, was affected afterwards and was consequently given a crossing keeper's job after a tussle between the Railway Department and the NUR.

At the inquest, the Coroner, Mr Stuart Rodgers, criticized the method of single line control and asked the Railway Superintendent, by then, Mr Hutchinson, if there was a system of control, adding that if there was a system it had lamentably broken down. Mr Hutchinson replied that there was a system and referred to the 1931 instructions of the, then, late Mr Alcock. He added that the instructions were not carried out properly. On being questioned by the Coroner about ensuring a more foolproof system of control, Mr Hutchinson outlined a proposed system which entailed the installation of extra telephones, sited at sections where there were lay-bys or passing loops.

Single Line Operation, Irwell Park-Latchford

A system, such as Mr Hutchinson had proposed, came into being on 24th June 1934 and incorporated a traffic position indicator board, as well as a recorded movement sheet. The control board layout is illustrated in the accompanying diagram; the line between Barton and Latchford being shown schematically, with circles on the line denoting sections and passing places; these are brass sockets on the actual board. The lettered and numbered circles in the rectangular box represent brass pegs, each denoting a different type of train. With a particular type of train in a section, the correct peg is placed in the socket for that section. Trains were only allowed to proceed as far as control directed. Even after a passing train had cleared a train standing in a loop, permission had again to be otained before it could be taken on to the single line. The signal at Barton was still used, however, the bridgemaster operating the signal only in the event that he could see the line was clear or if the aqueduct was not being turned. To alert the bridgemaster the shunter rang a bell in the bridge control room by pressing a button situated near the signal. One reason for this procedure, was the amount of water which cascaded from the aqueduct 'tank' as it was turned, for apart from the possibility of soaking the goods carried in the wagons, (it was sometimes quite a deluge), there was also the possibility of the water being dumped down the locomotive's chimney.

There were two other accidents recorded which, however, were not connected with single line working nor did they involve more than one engine. One involved locomotive 74, the 'posh engine'. In 1928 it was hit by its own train which had been left at the top of a gradient near Latchford Locks whilst a wagon of coal was being shunted into the locks siding. The second accident involved another Kitson but of unknown number, which, when approaching the entrance to Warrington Tube Works and hauling a heavy train, suddenly veered into the works yard, jumped the track and smashed through a wall. Luck again favoured the Manchester Ship Canal and no one was hurt. On examination, the points controlling entry into the works, which were supposed to have been locked, had been tampered with and were set to enter the works.

The control board schematic diagram at the control desk, Partington. The peg and socket board is illustrated in the text.

D. Thorpe

New Partington Locomotive Shed

However, there were some constructive, rather than destructive, events during the years under discussion. In May 1925 a contract was awarded for the construction of a new locomotive shed at Partington and this was in commission in September the same year. The original 3 engine timber built shed, which stood near Whitfield Street, was demolished the following year. In 1927 the new Dock Office on Trafford Road was completed and the Railway Superintendent and his staff were given a suite of offices in the new building.

Fixed Railway Bridge Replacement Programme

More constructive work followed with the replacement of the timber built railway bridges by concrete structures. By August 1928 the bridges at Rixton Junction and the River Glaze had been completed, whilst those at Boysnope Wharf, Moore Lane and Randles Sluices were still in the course of reconstruction. In 1937, at Mode Wheel workshops, a major expansion took place, which included a new boiler shop and paint shop, new offices, general stores and a canteen in one block, an extension to the wagon shop, and a new chain testing shop. Afterwards most of the major boiler repairs formerly carried out at Partington workshops were dealt with at Mode Wheel, a Partington boilermaker, Joe Morgan, moving to the new boiler shop to become its first foreman.

Manchester Ship Canal Railway Rule Book

On 1st July 1936 the new Manchester Ship Canal Railway Rule Book came into use, some of the changes being directly attributable to the bad accident at Barton in 1934. Later that month, in Trafford Park, a serious fire at the premises of Frank Rosser, timber importers, destroyed three timber yards and a train of 19 Manchester Ship Canal wagons. It happened

Kitson locomotive No. 74, the 'posh engine', after being hit by its own train at Latchford Locks. Beyond the tipped-up crane is the lock-keeper's office, and if the crane had not been stopped by a mooring bollard, the office and its occupants would have been crushed. The year was 1928.

Courtesy A. Day

Partington (north side) locomotive shed, circa 1962. Note the empty coal stage and the newer diesel fuel tank. The shed, which had a capacity for about eight engines, was built in 1925.

Courtesy H. Hudson

In 1927, the new dock office was opened and the Railway Department moved in, occupying most of the first floor. Note the tram lines and overhead wires serving the Trafford Rd. service of Salford Corporation Tramways.

M S C Co.

In 1928, during the replacement of many of the early railway bridges (most of them wooden) by reinforced concrete structures, the line at Moore Lane, near Warrington, was moved 25ft. nearer to the canal so as not to interfere with the old line during the reconstruction period.
Courtesy W. Forster

Moore Lane, Warrington, with a glimpse of the swing bridge seen through the support pillars of the concrete railway bridge built in 1928. This bridge replaced the earlier wooden structure which carried the single line running between Acton Grange and Wigg Works (Old Quay).
D. Thorpe

A rather charming little bridge is the one which carried the single line over the Glaze Brook near Cadishead.
D. Thorpe

Part of the course of the original River Irwell now forms the inlet leading to the former Manchester Corporation's Boysnape Wharf. The entrance is spanned by the single line bridge which is shown here.

D. Thorpe

A glimpse of the wooden railway bridge spanning the River Mersey as it leaves the Canal at Rixton Junction. This was one of those temporary structures erected by the contractor during the construction period, and was still in use until it was replaced in 1928 by the present concrete structure.

M S C Co.

Rixton Junction is not a railway junction but the point where the River Bollin enters and the River Mersey leaves the Canal. Spanning the course of the River Mersey is the longest of the concrete bridges carrying the Manchester Ship Canal Railway. It is a single track bridge, and replaced the early timber structure in 1928.

D. Thorpe

Randalls Sluices, spanned by the Pearson & Knowles iron road bridge. Also of interest is the simple timber structure spanning the sluice spillway, and carrying the single Manchester Ship Canal line which ran between Acton Grange and Wigg Works, Runcorn.

M S C Co.

Randalls Sluices spillway and the single track railway bridge which replaced the earlier timber structure in 1929. In December 1981, both sluices and railway bridge were dismantled, and the spillway was filled in. Immediately behind the camera are the waters of the River Mersey.

D. Thorpe

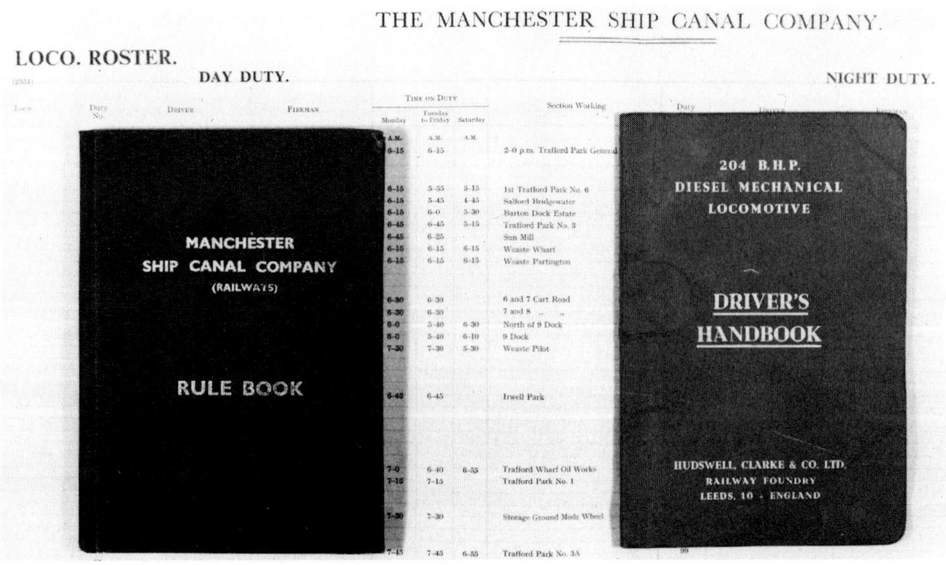

after a long hot spell of weather had made conditions particularly favourable for an accident of that sort to occur. This was only the third incident recorded, up to that date, in which Manchester Ship Canal wagons were destroyed by fire. The first incident happened in the L&YR siding during August 1911, only one wagon being lost. Three wagons were destroyed in the second incident which occurred in Trafford Park on 12th June 1919.

World War II

A much more terrible period of destruction was heralded on 3rd September 1939, for on that day Great Britain declared war on Germany. Unlike World War I, the possibility of a war had been recognized for many months, therefore, it came as no shock when it occurred. Again, unlike World War I, this war was to be fought over the heads of the civilian population as well as on the battlefields. Obviously the main civilian targets were the factories, docks and transport facilities. Manchester was a prime target, not only as an area with large production capability, but also as a strategically-placed port. During the three nights preceding Christmas Eve 1940, enemy bombers pounded the area with high explosive bombs, land mines and incendiaries. There were thousands of casualties and hundreds were killed. Houses, factories and transport facilities were wrecked but, miraculously, each morning the civilian population were back at their jobs even when the factories were roofless or worse. Trafford Park and Manchester Docks received a lot of damage but not enough to incapacitate them. On Trafford Wharf, No. 1 Grain elevator was burned to ashes during the night of 23rd December. It was a wooden structure and, after it was built in 1897, had been hailed as a 'most remarkable labour saving device'.

The railway itself sustained remarkably little damage, but 85 wagons were destroyed and many more damaged. One report tells of a Hunslet side tank engine, a 'Jazzer', losing its bunker after running over an unexploded bomb outside Eccles Oil Works. However, as the driver, Charlie Brady stated, 'only the cab windows were blown out, but there was certainly nothing more serious'. The engine involved was locomotive 15 *New York*. Its sister engine, locomotive 16 *Galveston* also received damage through shrapnel.

At Mode Wheel workshops a land mine blew off the entire north-light roof of the fitting machine shop. In the shop at the time were locomotives 51, 79, 38 and 65. Apart from locomotive 51, the engines were in the shops for twenty two weeks before they were able to work again. Locomotive 51 had received a heavy overhaul and was about to leave the shops when the air raid took place, but this engine was only delayed for seven days.

Another casualty was the Manchester Ship Canal Warehouse, near Vere street, which had been leased to the previously mentioned fruit importers W. W. Jones Dooly & Company Limited. The warehouse was completely destroyed and the firm was forced to look for new premises, subsequently ending up in Trafford Park. Strangely enough, over 40 years later, the area, which became the Resident Engineer's Department track maintenance depot, was still referred to as Dooly's Yard.

A new junction, Thornton North, was opened on 25th October 1940. It was situated at the western end of Stanlow and was made to ease the bottleneck of the single line connection at Ellesmere Port East. During its years of existence, Thornton North was to be opened and closed several times. After the war had ended the Manchester Ship Canal decided

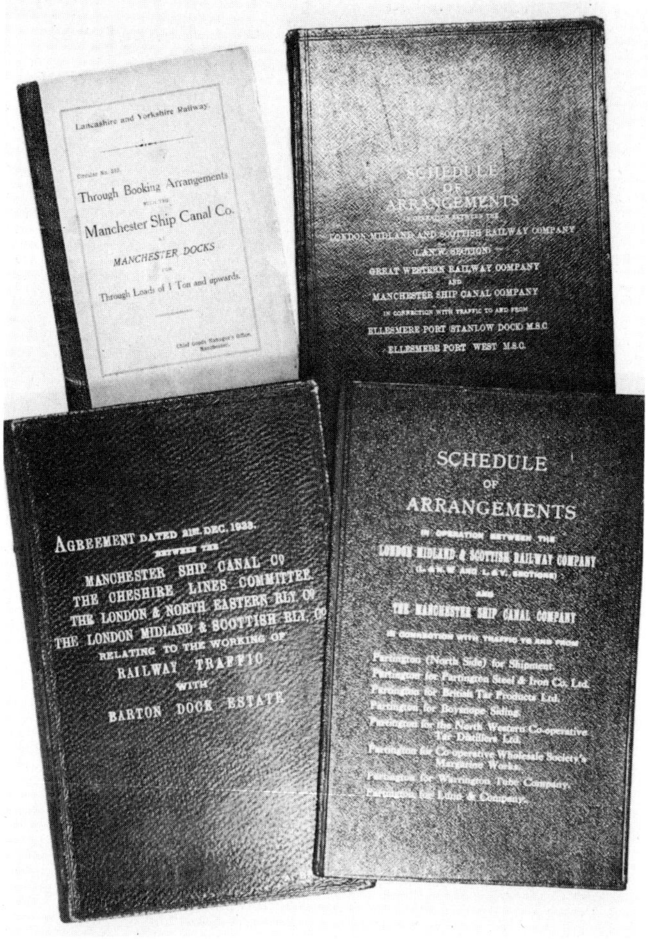

Various railway companies' agreements. These are but a few of the many which were entered into by the Company.

D. Thorpe

to close it, the date being 8th May 1946. At the request of BR the junction was reopened on 12th June 1950, but by 1980, it had fallen into disuse until finally, towards the end of 1980, it was considered closed but was still left connected.

Returning to the war period, on 4th November 1940, Mr Hutchinson issued a memorandum to the effect that due to the existing situation, it had been decided that he should take overall control of the railway in all sections. In explanation, the previous Railway Superintendents had confined their attentions to the running of the railway in the area between Manchester and Wigg Works, Old Quay, the operations at Runcorn and Ellesmere Port and Stanlow being organized by the local superintendents. Even so, the superintendents at Runcorn, Mr Blackley, and at Ellesmere Port, Mr Browning were to continue supervision to a certain level in their areas.

Replacement of Railway Swing Bridge, Salford

Another effect of the war situation was the increased traffic on the railway, especially that which poured in and out of Trafford Park, where most of the factories had changed over to war production. Movement of essential traffic was being hampered by the single track railway swing bridge at Salford, and eventually permission was granted by the Government for its replacement, the materials, especially steel, being strictly

Pomona Docks, Manchester, taken from the Ordsall side of the Canal. The railway into Pomona is fed from the right of the photograph. Clearly seen is Bridgewater Canal and, to the rear, lines and rolling stock of the Cheshire Lines Committee.

M S C Co.

Loading tanks into concrete barges for transhipment into ocean-going vessels. The photograph was taken on Salford Quay near No. 9 Dock in 1944.

M S C Co.

The loading of armoured half-track vehicles from Manchester Ship Canal Railway wagons near No. 9 Dock in 1944. The bogie bolsters were themselves ex-War Department stock from World War I.

M S C Co.

controlled, and a new double track bridge was constructed. On 2nd April 1943, Mr Hutchinson issued a memo stating that the new bridge was to be used after 11.00a.m. on 5th April and that use of Railway Swing Bridge No. 1 was to end as from the stated time. There were reports of a shift in the pivot support masonry in the period immediately before it was declared open, but no record of any such incident has been found.

An amusing story of the opening illustrates the rivalry which existed between footplatemen. As the time approached for the opening of the bridge, two locomotives drew to a stand, one on the Trafford Wharf side, and one on the Salford Yard side, each driver hoping to be the first to cross. As soon as the finger on their watches touched 11 o'clock the locomotives rushed on to the bridge. Alas for locomotive 24 *Philadelphia* the honour of being the first to cross went to locomotive 31 *Hamburg*.

Wartime Movements of Staff and Locomotives

With regard to staff, some men were taken into the armed forces and this added to an already difficult situation, but with regard to engines, during 1940 there were not enough of them. In 1938 two locomotives, Nos. 34 *Copenhagen* and 26 *Bristol* had been sold, leaving the Manchester Ship Canal with 72 locomotives, 7 being at Ellesmere Port-Stanlow, 2 being at Acton Grange, 2 at Runcorn, about 10 at Partington, and 2, the Mechanical Engineer's and the Resident Engineer's locomotives, which were not used for traffic purposes. This left Manchester with 49 engines. This was not enough to fill the needs of the Railway Department and extra engines were sought. The first engine obtained was a short tank Hudswell Clarke 0-6-0 of the same type operated by the Company. It was a second-hand engine by the name of *Jeannie*, and after entering Mode Wheel shops on 18th July 1940, was rebuilt, leaving the shops on 19th September. In the same month, the Transport Department of the Ministry of War transferred 8 Port of London locomotives to Manchester. London's docks were being reduced to rubble by German bombing, therefore the Port of London Authority engines were being sent to ports which were still fully operational. A further batch of locomotives, complete with crews, which were units of the United States Army Transportation Corps, arrived in 1942. As the locomotive strength increased the Engineer's Department were in a better position to loan engines to others in need. Subsequently locomotive 15 and locomotive 32 were sent to work for Lever Bros., at Port Sunlight, Cheshire. They were sent to Port Sunlight at the end of December 1942, locomotive 15 working at Bromborough Dock until it was returned in August 1944, and locomotive 32 working at the top siding, Port Sunlight, until

Railway swing bridge No. 2, again built by the Leeds firm of Messrs Butler & Co., was a double track structure weighing around 300 tons, against the 65 tons of No. 1 bridge. It came into use at 11a.m. on 5th April 1943. Although it was built to speed up the flow of wartime traffic, which was being hampered by the earlier single line bridge, two years elapsed between its sanctioning and its completion.

D. Thorpe

An aerial view of the Salford group of docks showing (from lower middle) Nos. 6, 7, 8 and 9 Docks. It is also possible to see No. 1 Grid, New Barns Junction between Nos. 8 and 9 Docks (top) and, in the lower right corner, the line from Salford yard passing under Trafford Road to reach Ordsall yard. In the lower left are the railway swing bridges, the smaller of the two being the redundant No. 1 bridge.

M S C Co.

Locomotive No. 34 *Copenhagen* with its crew, pictured on the incoming line to Runcorn docks. The line, having been laid on banked earth, was known locally as 'T tippin'. Leaning on the buffers, in characteristic pose, is the Driver, James Joyce, and the photograph dates from about 1935.

Courtesy H. Joyce

In the early part of World War II, the Railway Department was inundated with traffic and the demands on the Department were very heavy. One of the engines brought in to ease the load was a second-hand Hudswell Clarke short tank, *Jeanie*, supplied when new in 1904 to Thomas Wrigley, the railway contractor who named many of his engines with ladies' names beginning with 'J'. *Jeanie* became the second MSC No. 34 but retained her nameplates on the tank sides, unlike the other named engines which had their plates removed in 1915 and had only one mounted on the rear cab plate.

K. J. Cooper, Courtesy J. A. Peden

An aerial view taken over Partington coaling basin. In the background is the works of The Lancashire Steel Corporation (formerly The Partington Steel & Iron Co. Ltd.).

Airviews (M/cr) Ltd., Manchester

An aerial view of part of Runcorn Docks. The sweeping curve of the railway is clearly seen and in the centre of the picture is Percival Rd. Power-Station.

M S C Co.

its return to Manchester in February 1945. All the Port of London Authority engines had been sent back to London by February 1946.

With the arrival of the US Army locomotives and crews, especially the crews, the Manchester Ship Canal footplatemen were introduced to a new world of railway jargon and methods. The locomotive shed became the roundhouse, and the line became the railroad. Drivers were engineers, firemen were stokers whilst the shunter became a rope runner.

Day-to-day servicing such as oiling round, replenishing oil bottles, and cleaning ash pans was done by a hostler. All the US personnel wore gloves and apparently, if one witness is to be believed, were never seen without a cigar between their teeth. Their casual and seemingly happy-go-lucky approach to the job was in complete contrast to that of their more regulation-conscious British counterpart. Before the end of the war all the US Army locomotives had been removed, reportedly to Europe. Members of the Royal Engineers also worked on the Manchester Ship Canal Railway during the war period but no War Department locomotives were involved.

During the war years, the Manchester Ship Canal Railway worked under tremendous pressure, men and machinery toiling every hour of the day and night. Almost a hundred locomotive shifts each week were being averaged. In 1943 railway tonnage reached an all-time record with almost 9,000,000 tons being moved, a figure which surely will never be reached again.

Runcorn, however, had rather a difficult time, with delays occurring, in some measure, due to the compactness of the docks. Another drawback was the use of a horse for quayside shunting. Only one wagon at a time could be taken by the horse. A request for a tractor shunter of the Muir Hill type was turned down by Mr Hutchinson, who pointed out that the soft earth of certain quays was not suitable for this type of machine and, furthermore, that there were already two locomotives on call at Runcorn which was sufficient for their needs.

To ease the situation with regard to the increased output from the (I. C. I.) Salt Unions No. 11 Works, a new line was put in on behalf of the Government, which connected directly to the LMS lines. Effectively, this took traffic off the Manchester Ship Canal Railway at Runcorn and, with the ending of the war, resulted in a wrangle with I. C. I. over loss of revenue. Unfortunately for the Company, the traffic was never regained.

End of State of War

On 8th August 1945, Japan, Germany's last ally, capitulated and the war ended. With the ending of hostilities the railway faced further problems, not the least being the poor condition of the track and equipment. The plumetting tonnage figures, as war production ended, was to be expected, of course, but it was only a comparatively short time before industry once again began turning out peacetime products.

Industrial Development 1925-45

During the 1925-45 period, industrial growth along the Canal had been rather desultory with only eight firms establishing their premises on the Company's estate. On the 'island', near Wigg Works, a chemical plant was built, in 1926, for the Chemical & Metallurgical Corporation. At Ellesmere Port the papermakers, Bowater Limited had established a mill alongside the Canal in 1928, and a year later Stanlow attracted an oil storage firm, Berry Wiggins Ltd. Another oil company, Lobitos Oil Field Ltd., settled at Stanlow in 1934. There were three firms established in the Manchester area, R. &. W. Paul Ltd., corn millers, Ordsall, 1932, the Kellogg Company (Great Britain) 1938, alongside the Bridgewater Canal on Barton Dock Estate, and also on the estate, in 1940/1, Reif & Son Ltd., timber importers. In 1943, at the wartime Government's request, a new factory was built at Latchford for the British Aluminium Company Ltd.

The Kellogg factory received grain and most of its coal by barges using the Bridgewater Canal, but in the early years most of the output went by rail. Reifs received most of its timber by rail direct from the quayside. British Aluminium and the Chemical & Metallurgical, later Kemet, works both received and despatched by rail. Bowater rarely used the railway for anything other than coal, but this commodity was delivered daily to the plant in very substantial quantities. Each of the remaining firms used the railway extensively.

In the twenty years leading up to 1945, there had been ten years of trade recession (1926-36) and almost six years of war. During the slump, or depression, period the railway managed to justify itself, gaining some traffic, but losing other traffic as firms closed down and as road transport gained ground. But for the emergence of the Stanlow development and, to some extent, the Barton Dock Estate development, the railway tonnage figures would have been a lot lower than they were. It is ironic that just as the economic situation was showing signs of improvement with industry becoming more confident, a state of war was experienced for the second time in 23 years. For the railway, war imposed a tremendous burden on men, machines and track system. It also illustrated just how invaluable a railway is when speedy, mass movement of essential freight is required.

A crowded Trafford Wharf, circa 1930, with locomotive No. 70 with its train in the foreground. In the distance, the tall structure is part of Grain Elevator No. 1, which was burnt down in the early part of World War II.

M S C Co.

Locomotive No. 39 *Sydney*, with a train of coaches for the Palestine Railways. The photograph was taken about 1935 on No. 1 Grid.

M S C Co.

Various tallies used for pay and equipment.

Chapter Four
Winning Through War, Losing in Peace (45-84)

The situation after the war had ended was, as was to be expected, a time of anticlimax with a lull in production as industry made the transition from the needs of war to those of peace. During 1946, railway tonnage plumetted to a yearly figure of 5,714,855, but the following year the total had risen sharply to reach 6,190,986. The recovery could be judged by the average monthly locomotive shifts that had been worked. For 1944 the figure was 91 locomotive shifts, by 1946 the figure had dropped to 70 shifts, but then the figures began to rise again. The year 1947 averaged 73 shifts each month, 1948 had 79 shifts and 1949, the peak post-war year, showed an average of 80 locomotive shifts worked. Unfortunately, the following year showed a slight drop, which is now seen as the beginning of the decline in the use of the railway.

Thereafter there were, of course, yearly fluctuations and the overall picture did not appear a gloomy one. The shipping tonnage continued to rise over a period of years and in the main the railway was extremely busy; in fact to work on the railway was considered to have a job with a future.

Renewal of Track System

Whilst the increased tonnage in 1947 was obviously welcomed by the Railway Department, for the Resident Engineer's Department and the Mechanical Engineer's Department it added to the problems caused by the war worn track and equipment. In the immediate post-war years, materials, including railway track, was in very short supply, therefore, the renovation and upgrading of the system had to be accomplished over a period of years, most of the track used being ex-BR.

In the early years practically the whole system was laid in 56lb. and 60lb. flat-bottomed (Vignoles) rail, presumably that which was taken up from the contractor's railway. In many of the busier sections the rails were secured by wooden clamps instead of spikes. By the mid-1920s the through route and the more important sidings were laid in 75lb. bullhead rail. With the upgrading of the system after 1946, the rail used was 95lb. bullhead for the busiest lines with some 75lb. rail in the less important sections. There was, however, some 60lb. flat-

Relaying railways on No. 8 Pier, Salford Docks, on 20th May 1963. The lines have been levelled and are in the process of being encased in concrete. Jobs of this nature were usually let out to contractors and this contract, No 617, was no exception, the firm handling the job being Monk Ltd., Warrington.

M S C Co.

bottomed rail still down in certain timber fields and sidings in the late 1950s. After 1953 when the effects of Beeching's axe began to be seen, the Manchester Ship Canal purchased a great deal of track, mostly bullhead, but also including 110lb. and 113lb. flat-bottomed rail. One batch came from Penge Tunnel and others came from the Pennine routes. A batch of 1,000 tons was shipped direct from the Isle of Wight to Manchester Docks. Although the rail could be had comparatively cheaply, it was still a costly process, although the post-war programme was partly paid for by the Government in war reparations.

During the relaying period the platelayers received mechanical aid in the form of a steam crane. There were also petrol-engined saws and drills, but it was not until the main task of upgrading the track had been completed that the more specialist machines were employed. In 1964 a 'Mattisa' ballast tamping machine was delivered and was put to work at Salford, and later the same year a 'Robel' motorized gangers' trolley was purchased.

A Hudswell Clarke side tank locomotive, fitted with a snowplough, at No. 29 Bonded Stores, Eccles during the bad winter of 1947.
J. Steggles

Port of London Authority engine PLA 39, built by Andrew Barclay & Sons in 1911. This locomotive, which is seen on home ground, was one of eight PLA engines which were loaned to the Manchester Ship Canal during World War II, two others of the same type and five Hudswell Clarke 0-6-0 outside cylinder side tanks making up the total.
S. W. Baker, Courtesy R. C. Riley

Port of London Authority Locmotives, and Manchester Ship Canal Locomotive Developments

By the end of January 1946 all the Port of London Authority engines had been returned to London after overhaul at Mode Wheel shops. In April that year the Manchester Ship Canal locomotive rebuild programme, which had been started in 1937, and slowed by the war, started again when locomotive 39 entered Mode Wheel fitting shop. It was a slack period briefly for the railway department and it was not making such heavy demands on the Locomotive Department, therefore, the loss of the Port of London Authority engines and those Manchester Ship Canal engines due for overhaul or rebuild, was not felt quite so badly. As the year went on, however, and the railway tonnage began to ascend, the demands of the railway department became heavier. Consequently the Mechanical Engineer decided to obtain new locomotives, but of a more powerful type. The year 1947 saw the arrival of the first of three ex-War Department engines. Later, an order was placed for the building of three similar, but heavier engines, these being delivered in 1954. The year 1958 marked the first year of dieselization with the arrival of the diesel-electric engines from the makers of the last steam locomotives, Hudswell Clarke & Company. The dieselization programme was completed in 1966.

Other signs of the Company's faith in the future of its railway were the erection of new workshops at Partington in 1964, and at Ellesmere Port in 1970, and of new railwaymens' accommodation at Salford in 1964. The accommodation, which cost £10,000 was used by Inspectors, foremen, clerks and shunters, also sheeters and platelayers.

Industrial Developments 1945-1979

Industrial development proceeded apace. Each week, large amounts of scrap aluminium were being taken by British Aluminium as it switched over to peacetime production. In 1946 a former London export company, Viarens Ltd., which usually exported hydroelectric machinery and sometimes coaching stock, mainly to Portugal, established premises at Pomona. The following year at Barton Dock Estate, Courtaulds Ltd. started work on a new plant for the production of sulphuric acid. The plant was actually developed in stages over a span of 24 years. At Warrington in 1948 another large chemical plant was started. It was built for Laporte Industries Ltd., and Baronet Works, as the plant was named, was sited on the 'island' near Walton Old Junction (Manchester Ship Canal Acton Grange).

As the works came into production in 1950, an American company, Cabot Carbon (UK) was starting a factory on the southern side of the Helsby-Hooton (BR) line at Stanlow. The raw material, carbon black, and the firms fuel requirements were brought in by rail. A less messy cargo was delivered to another new firm, Morgan Wallworks Ltd., timber importers,

A fine picture of locomotive No. 90, the last steam engine to be purchased by the Manchester Ship Canal Co., with driver Albert Hulme in a very cheerful mood. Built by Hudswell Clarke & Co. in 1954 (Works No. 1867), No. 90 and the other two members of the class arrived too late in the day to survive any length of time and, in fact, were put to the torch only ten years after being built. The photograph was taken on 18th May 1954, on No. 1 Grid facing the rear of the warehouses of No. 8 Dock.

A. Appleton

Hudswell Clarke, diesel-electric No. 4001, (Works No. D1075) of 1959, passes the dock office at Salford early in 1959. One of the most pleasant surprises about the locomotive when it arrived was the reintroduction of names. The name *Alnwych Castle* was given to this particular locomotive.

Courtesy R. N. Redman

The new Partington Workshop in the course of construction during 1964. To the left of the picture is the north side locomotive shed, whilst at the rear are the chimneys of Irlam Steelworks.

W. Williams

Railwaymen's accommodation near No. 2 Grain Elevator, on No. 9 Dock. Built in 1964 at a cost of £10,000, the building housed members of the Railway Department, including Inspectors, foremen, clerks and shunters. It also housed sheeters and members of the Civil Engineer's Dept., such as platelayers, etc.

M S C Co.

The Shunters' cabins at No. 9 Dock (North). These are typical of those which were used all over the Canal's railway system.

Courtesy H. Hudson

A view of the Manchester Ship Canal's Ellesmere Port to Eastham line, which was built in 1967. The lines on the right of the photograph are those of Bowaters Ltd.
M S C Co.

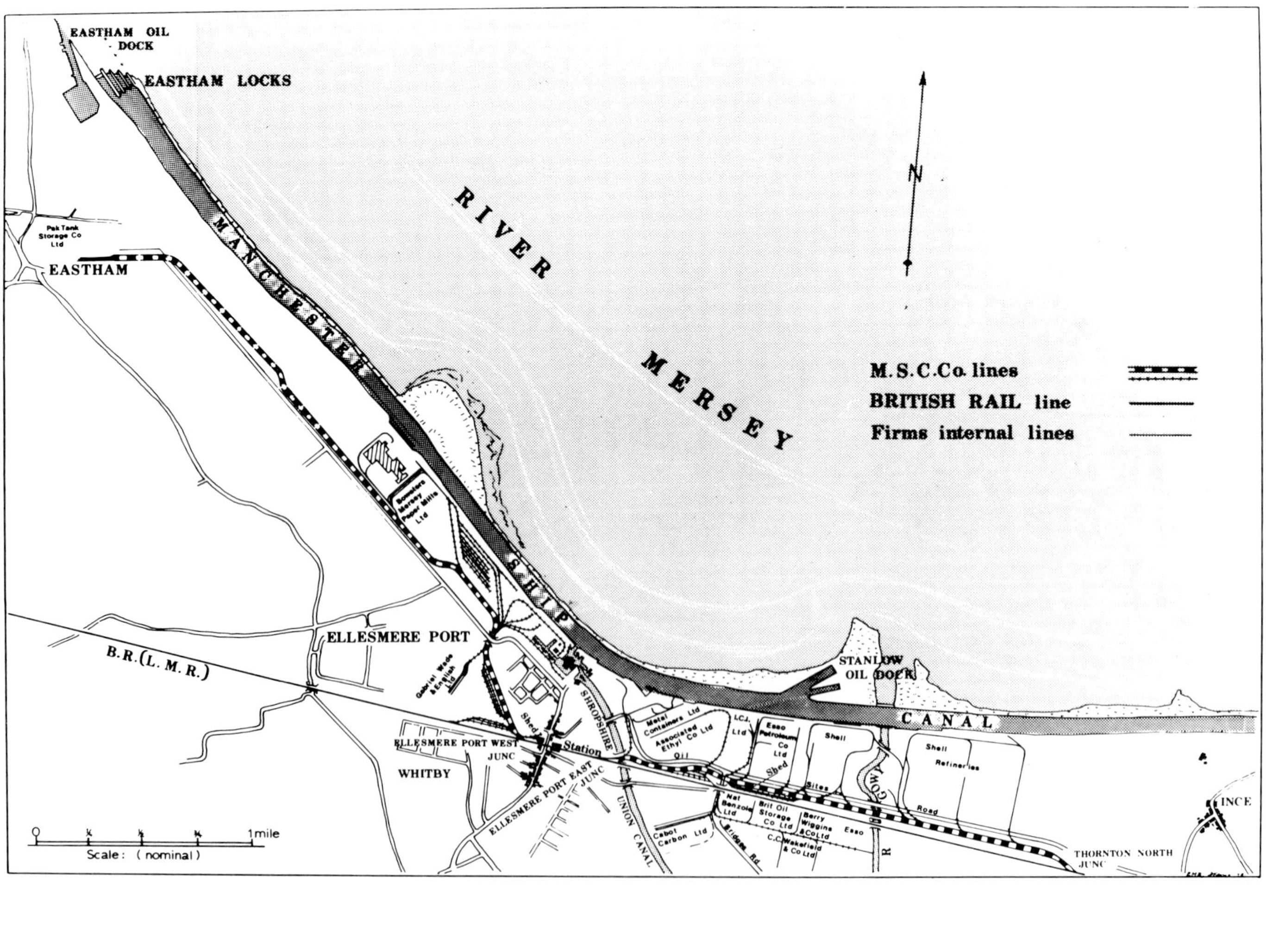

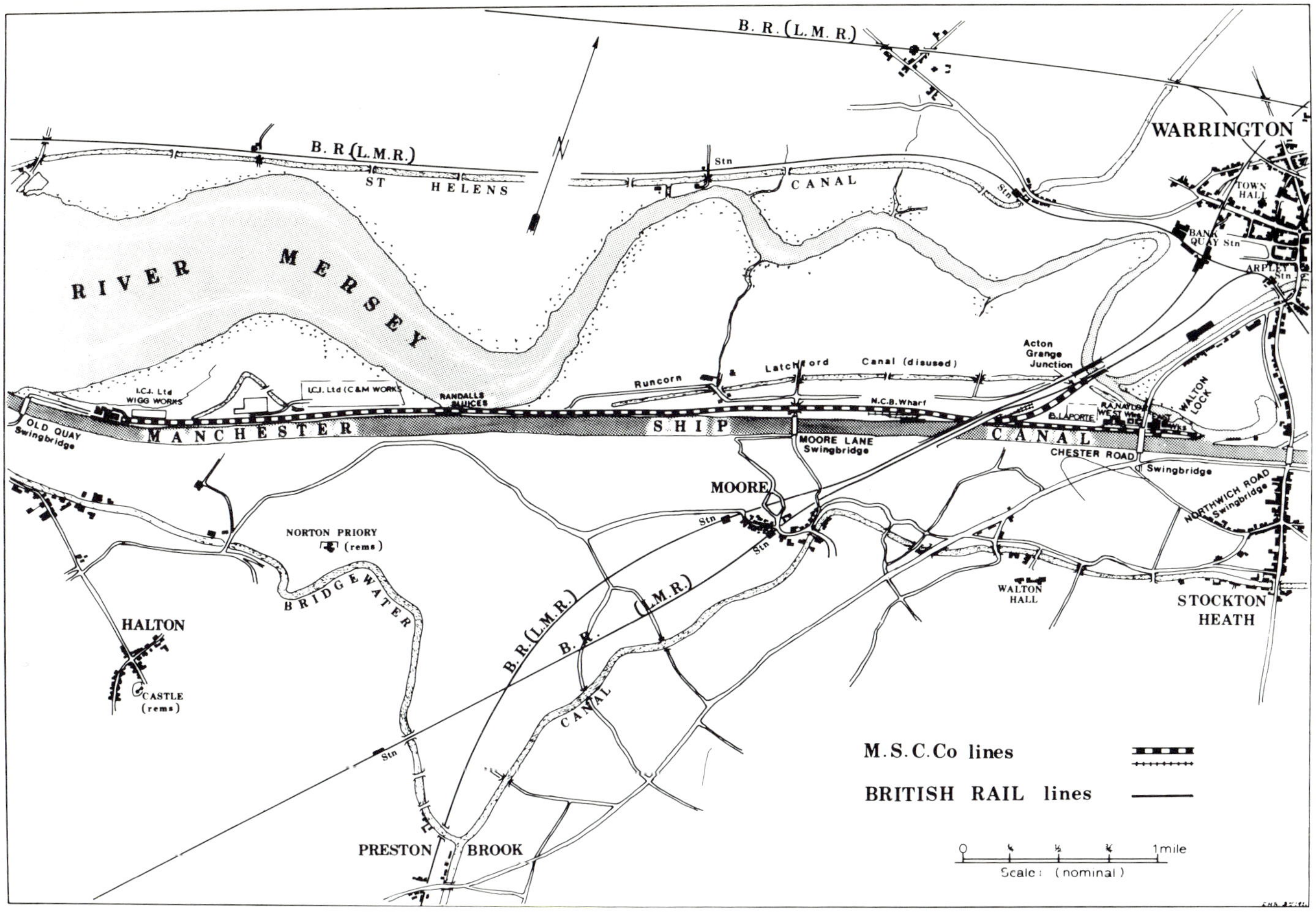
B. R. (L. M. R.)
WARRINGTON
TOWN HALL
BANK QUAY Stn
ARPLEY Stn
B. R. (L.M.R.)
ST HELENS CANAL
Stn
RIVER MERSEY
Acton Grange Junction
Runcorn & Latchford Canal (disused)
I.C.I. Ltd WIGG WORKS
I.C.I. Ltd (C&M WORKS)
RANDALLS SLUICES
N.C.B. Wharf
WALTON LOCK
OLD QUAY Swingbridge
MANCHESTER SHIP CANAL
MOORE LANE Swingbridge
CHESTER ROAD Swingbridge
NORTHWICH ROAD Swingbridge
MOORE
NORTON PRIORY (rems)
HALTON
CASTLE (rems)
BRIDGEWATER CANAL
B.R. (L.M.R.)
B.R. (L.M.R.)
WALTON HALL
STOCKTON HEATH
M.S.C.Co lines
BRITISH RAIL lines
PRESTON BROOK
0 ¼ ½ ¾ 1mile
Scale: (nominal)

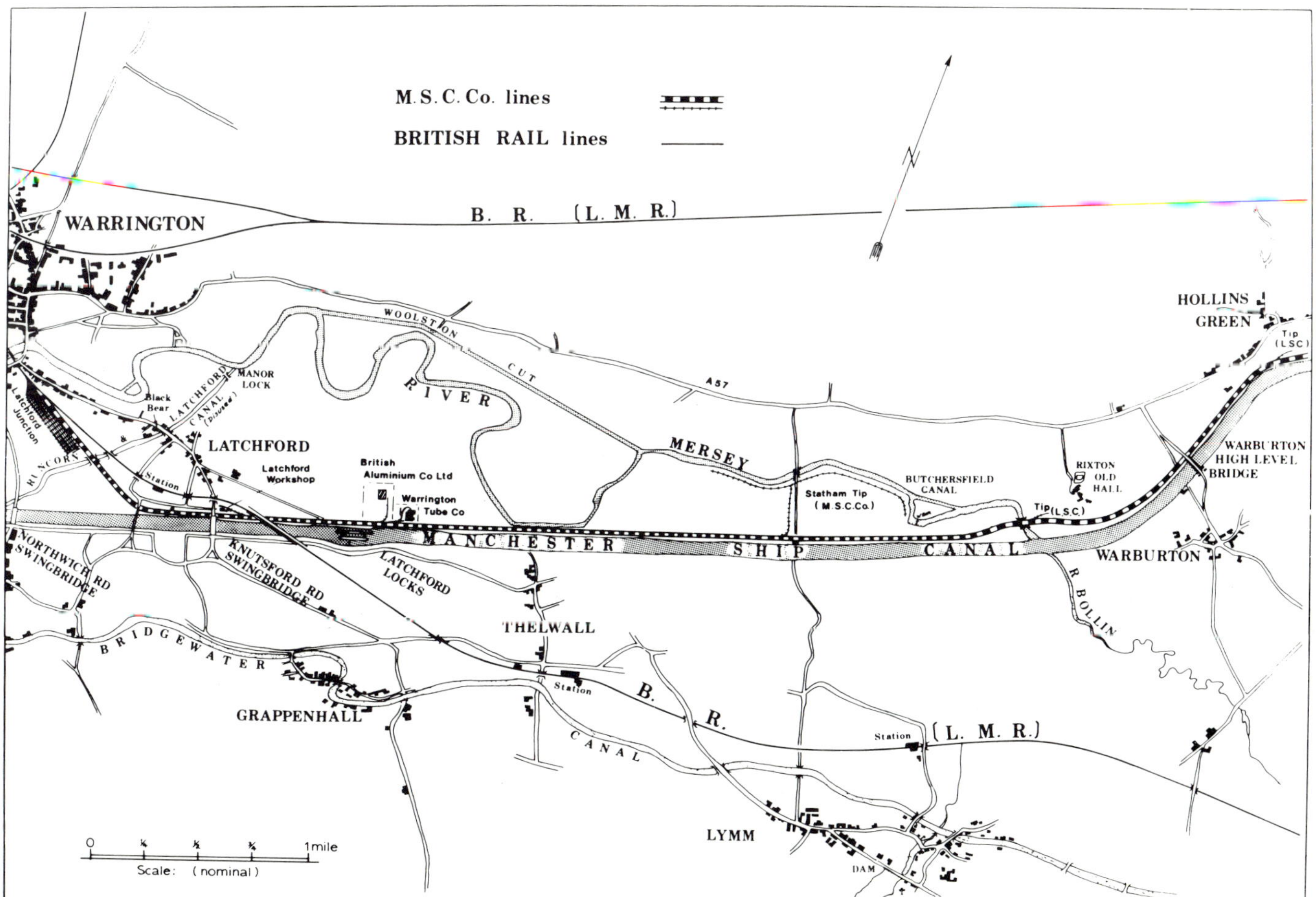
M.S.C. Co. lines
BRITISH RAIL lines
WARRINGTON
B. R. (L. M. R.)
HOLLINS GREEN
Tip (LSC)
WOOLSTON CUT
RIVER MERSEY
A57
MANOR LOCK
Black Bear
LATCHFORD CANAL (Disused)
Latchford Junction
RUNCORN
LATCHFORD
Station
Latchford Workshop
British Aluminium Co Ltd
Warrington Tube Co
Statham Tip (M.S.C.Co.)
BUTCHERSFIELD CANAL
RIXTON OLD HALL
WARBURTON HIGH LEVEL BRIDGE
Tip (L.S.C.)
MANCHESTER SHIP CANAL
WARBURTON
NORTHWICH RD SWINGBRIDGE
KNUTSFORD RD SWINGBRIDGE
LATCHFORD LOCKS
THELWALL
R BOLLIN
BRIDGEWATER CANAL
GRAPPENHALL
Station
B. R. (L. M. R.)
Station
LYMM
DAM
0 ¼ ½ ¾ 1mile
Scale: (nominal)

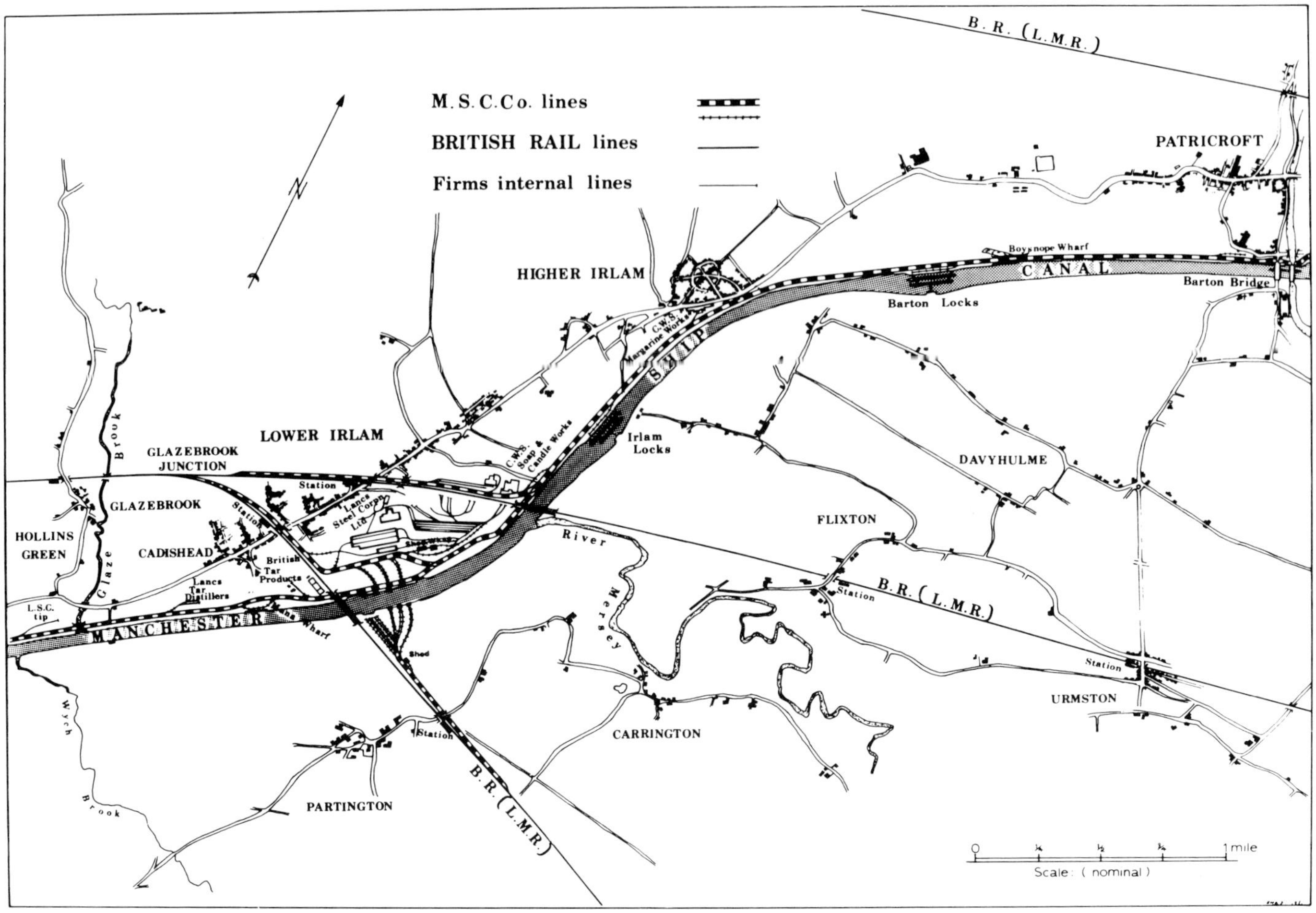
M.S.C.Co. lines
BRITISH RAIL lines
Firms internal lines
B.R. (L.M.R.)
PATRICROFT
Boysnope Wharf
CANAL
Barton Bridge
Barton Locks
HIGHER IRLAM
C.W.S. Margarine Works
SHIP
C.W.S. Soap & Candle Works
Irlam Locks
LOWER IRLAM
GLAZEBROOK JUNCTION
Glaze Brook
GLAZEBROOK
Station
Lancs Steel Corpn Ltd
HOLLINS GREEN
CADISHEAD
British Tar Products
Lancs Tar Distillers
L.S.C. tip
MANCHESTER
Wharf
Shed
River Mersey
DAVYHULME
FLIXTON
URMSTON
CARRINGTON
PARTINGTON
Wych Brook
Scale: (nominal)
1 mile
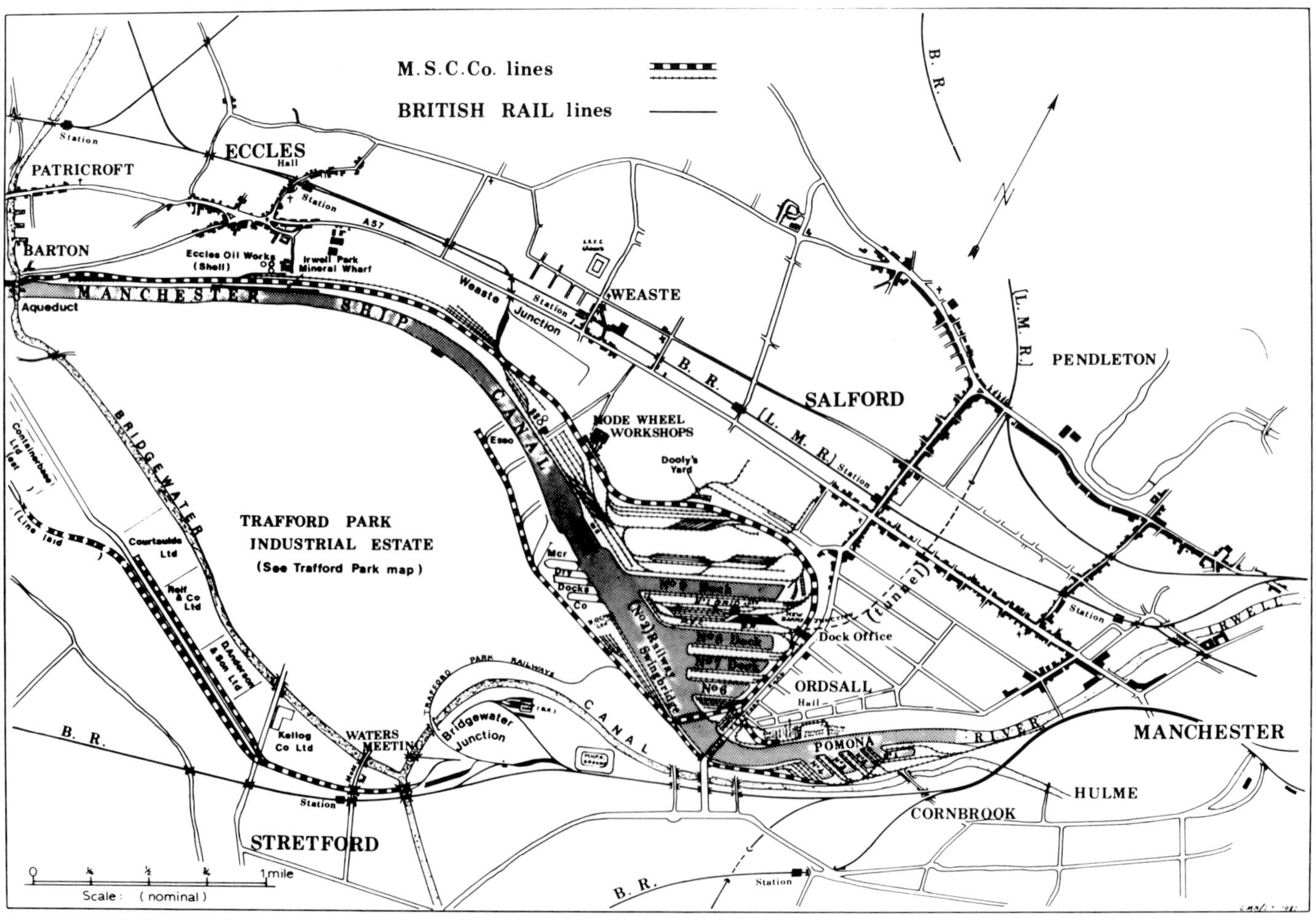
M.S.C.Co. lines
BRITISH RAIL lines
B.R.
B.R. (L.M.R.)
PATRICROFT
ECCLES
Hall
Station
A57
BARTON
Eccles Oil Works (Shell)
Irwell Park Mineral Wharf
Aqueduct
MANCHESTER SHIP CANAL
Weaste Junction
WEASTE
PENDLETON
SALFORD
MODE WHEEL WORKSHOPS
Esso
Dooly's Yard
BRIDGEWATER
TRAFFORD PARK INDUSTRIAL ESTATE
(See Trafford Park map)
Courtaulds Ltd
Containerbase Ltd
(Line laid)
Mcr Docks Co
No 9 Dock
No 8 Dock
No 7 Dock
No 6
Railway Swingbridge
Dock Office
(tunnel)
ORDSALL
Hall
IRWELL
RIVER
MANCHESTER
POMONA
HULME
CORNBROOK
TRAFFORD PARK RAILWAYS
CANAL
Bridgewater Junction
WATERS MEETING
Kellog Co Ltd
STRETFORD
Scale: (nominal)
1 mile

after its premises were established at Irlam on 21st June 1955. It was another ten years before another firm settled on the Canal estate. In 1965 Lancashire Tar Distillers put a light oil terminal in at Weaste. The terminal was built within a wired compound, the oils being pumped to the firm's tanks at Borough Road, Weaste.

In an era when the usual news was of railways being lifted it came as a pleasant change to witness the laying of a new line, 2 miles with a 280 yard loop, between Ellesmere Port and Eastham. The line was put in to serve the liquid storage tanks of Pak-Tank Ltd., later Unitank, and was completed in 1967. Another firm Pan-Ocean Terminals Ltd., established a depot nearby in 1972, but did not start to use the railway until 1979. The phenomena of laying new lines was to be repeated in 1968 when a 1,600 yard length of double track was laid at Barton Dock Estate leading to the premises of Containerbase Ltd. As can be imagined, Containerbase was a depot for the transhipment of containers between lorries and railway wagons.

The loading of the Southampton Maritime train at the container base at Barton Dock Estate. The man at ground level is in radio contact with the container gantry operator and gives the necessary guidance. Alongside the gantry is the vehicle used to bring and remove trailers, and this is known as a 'tug'. The container gantry is a remarkable machine and is fully steerable on all wheels.

D. Thorpe

Early Signs in the Declining Use of the Railway

With the scale on which improvements and extensions to the railway system was being carried out, coupled with the settling of firms on to the Manchester Ship Canal estate, it comes as something of a shock to realize that it was all happening with a background of falling traffic figures, rail closures and staff redundancies.

Runcorn

Runcorn was the first section to dispense with the railway, the official date of closure being 31st December 1951. Even though it was the oldest established section of the Manchester Ship Canal Railway, the Bridgewater Department's Railway had not been organized to its full potential for many years for various reasons, not the least being that Bridgewater Department operated a large road transport fleet.

The locomotive fleet comprised two Hudswell Clarke saddle tank 0-6-0s, locomotives 75 and 78, moved to Runcorn in 1938, and a Peckett, locomotive 25 which had been transferred to Runcorn on 11th March 1949. A sum of £1,000 was debited against Bridgewater Department. Since the arrival of locomotive 25, locomotive 75 had been up for sale, but was in poor condition and not attracting any buyers at a selling price of £500. The Mechanical Engineer's Department was not prepared to spend an estimated £1,200 on the overhaul of the engine and it was subsequently scrapped. the sister engine locomotive 78 with locomotive 25 was returned to Manchester, and after being overhauled both were put to work at Salford. The remaining wagons at Runcorn were loaded on to barges and sent back to Manchester and, as was the usual practice when equipment was transferred between departments, a charge was made against the Mechanical Engineer's Department. A terse note from the Mechanical Engineer reminded the Superintendent that when the wagons, 24 in number, were sent to Runcorn in March 1949 no charge was made for their transport, therefore, charges should be waived, and the Superintendent agreed to this.

Irlam Steelworks Railway Bridge

A further eleven and a half years were to pass before the next closure, but during that period the Canal was to witness the opening of a new line which robbed the railway of a great deal of traffic. The new line was built for the British Steel Corporation and connected its Irlam Steelworks direct to the BR system near Irlam Station. The first inkling of what was

A container train leaves the container base at Barton Dock Estate on 12th June 1977. The train is usually handled by one of the Hudswell Clarke diesel-electric engines. On this occasion it was the turn of No. 4002.

D. Thorpe

With the closure of the railway at Runcorn Docks, all three locomotives were returned to Manchester. No. 25 was put to use at Salford but No. 75, being in poor condition, was subsequently scrapped. No. 78, the sister engine to No. 75, was overhauled at Mode Wheel Shops and afterwards repainted, this time in the new MSC livery of light grey, edged with black and lined with white. The engine is shown posed near the oil stores in Mode Wheel yard.

J. Steggles, Courtesy W. Clayton

Ex-MSC Peckett locomotive No. 11 *Alexandria* is pictured at the Esso Company's Mode Wheel premises in 1955. It is said that when Esso made enquiries concerning the Peckett, the firm's engineer thought they were getting the younger locomotive No. 25. Whether there is any truth in the story is hard to ascertain. What is known about the engine is that the boiler developed a leak, which progressed to a point where it would not hold the pressure, and after the MSC personnel (the author) had examined the fault, Esso had to withdraw No. 11 from traffic. The engine was scrapped about August 1958.

A. C. Sterndale

Spanning Liverpool Rd., Irlam (A57) is the railway bridge carrying the lines of the British Steel Corporation from its Irlam Steelworks to the BR metals near Irlam Station. Behind the girder bridge is the BR brick-built bridge constructed to carry the deviated lines of the Cheshire Lines Committee during the Canal construction period. At one time, the bridge carrying the MSC (ex-CLC) lines over the road lay on the far side of the brick arch. Before the Canal was constructed the road did not dip to go under the CLC lines, but went over the railway; the bridge being constructed and the road lowered by the Canal contractors.

D. Thorpe

to come occurred in January 1956, when the British Steel Corporation Board granted permission for money to be spent on providing the BR/British Steel Corporation link, a project which included the construction of a bridge spanning the A57 Liverpool Road at Irlam. The cost involved was to be in excess of £1 million. An Act was petitioned for, which of course, the Manchester Ship Canal Company opposed, but with no success. Eventually the work started and on a Saturday evening, 24th March 1957, the A57 at that point was closed temporarily to traffic whilst the spans were lifted into position. The bridge is situated on the westerly side of the BR (ex-Cheshire Lines Committee) bridge.

Although the bulk of the traffic was lost to the Manchester Ship Canal Railway, some traffic, mostly wire in coils, was retained, this being taken to Irwell Park wharf. This traffic finally ceased in 1972. By that time the steelworks, itself, was facing closure, the announcement being made in April the previous year, and the final shutdown happened early in 1979.

Further Railway Junction Closures

Returning again to rail closures, on 15th June 1963, New Barns Junction was closed, and the following year saw Irlam Junction terminated. Actually, from the Company's point of view, Irlam Junction had never figured as part of the system, as it had been used solely by the CWS from 1897 until about 1960. A regular vehicle on the line was the old ex-MR six-wheeled composite which was hauled by a Peckett between Irlam Station and the CWS Soap Works each morning and evening carrying CWS office staff. For the Manchester Ship

The CWS Soap & Candle Works locomotive in the works.
I Higginbotham

Canal, the cost of maintaining the line and the three bridges over which it ran could not be justified after the CWS ceased to use it. On the weekend of 4th & 5th December 1965 the plate girder bridge, over the A57 near Irlam Station, was dismantled by A. C. Brew & Co. A second bridge, a lattice girder structure (ex-Cheshire Lines Committee) spanning the dry bed of the Old River (Irwell) and a small plate girder bridge was dismantled shortly after by the same firm. It is interesting to note that for some unknown reason the line has always been known to the Manchester Ship Canal men as Queens Road. Partington (southside) Junction was the next to be closed, its use ceasing after 31st December 1966. All coaling operations were concentrated on the north side.

The section between Partington and Latchford suffered an early demise because of the continued landslips which, on several occasions, left the line dangling in mid-air. The landslips occurred on the north bank of the Canal at a point where the Thelwall Viaduct carrying the M6 over the Canal now stands, and was caused by the saturated state of the ground in that section. The area, was, in fact, a deposit ground for dredging spoil, and the trouble arose because the water was not draining away as fast as had been estimated. The first of the landslips occurred in December 1964 and as time went by they became increasingly serious, the line having to be moved further away from the canal each time. Eventually, in May 1967, after a particularly bad slip it was decided that in view of the diminished traffic using the line, the expense of making it safe could not be justified and the section was effectively closed. This obviously ended the pay coach run to Latchford and from that time, until it ceased to run altogether, the pay coach only went as far as Partington.

The decision to stop using the line virtually isolated the Latchford section and on 1st April the following year this too was closed down. Both British Aluminium and the Warrington Tube Company had no option but to accept the situation, and consequently they dispensed with their own railway equipment, which, in the case of British aluminium, included a small diesel locomotive, which was formerly used for internal shunting. Actually traffic in the section had fallen off over the years, the Warrington Tube Company producing very little of its special type of tubing. The work carried out in the factory was mainly done on jobs requiring ready made tubes. As it was, the firm went into liquidation in 1970 and the premises were eventually taken over by Country Apex Ltd., an engineering firm. There were no plans for the new owners to use the railway and the section remains closed.

On 7th October 1968 Weaste Junction was closed to general traffic, a single line being left to carry the Lancashire Tar Distillers oil traffic, which left only Bridgewater Junction to handle general traffic at Manchester. This was the measure of how bad things were, for Weaste Junction had been one of the busiest of the Manchester Ship Canal main line connections.

Acton Grange was the next casualty in the process of contraction, for traffic had been dwindling steadily over the years. The timber importers, R. A. Naylor & Sons, ceased using the railway in the mid-1960s, all timber being received and despatched by road. The National Coal Board wharf at Acton Grange closed down in February 1963. Although the site was taken over in 1967 by Dupont (UK) for the building of a warehouse, which came into use in 1968, the firm did not have a rail connection. Incidently, when the site was purchased by Dupont, the Manchester Ship Canal line ran through it and as part of the agreement, the new owners had a line laid, at their own expense, further to the north of the area. The old line had to be kept open whilst the new one was being constructed due to the traffic being taken to and from Wigg Works which, on some days, amounted to two trains, mostly of tanks. At a later date Dupont put in level crossing lights to control traffic using the approach road to its premises, as the Manchester Ship Canal line crossed the road

A timber train, headed by diesel-electric locomotive No. 4002 *Arundel Castle*, passes, on 27th July 1966, below the gated end of the Bridgewater Canal at Barton. The aqueduct (unseen) is in a position which allows passage for vessels using the Ship Canal.

I Higginbotham

The deserted BR sidings at Latchford after the closure of Latchford (Black Bear) Junction.

D. Thorpe

A tandem unit at work hauling a train of fourteen 100 tonne tanks (empty) out of the Lancashire Tar Distillers compound, towards Weaste Junction. The photograph, taken on 5th May 1982, shows Rolls-Royce Sentinel 0-4-0s Nos. DH24 and DH26 forming the unit. A third 230hp locomotive, No. DH23, is the other member of the class which can be coupled to either of the two shown to form a tandem unit. The driver can be seen in the second locomotive.

D. Thorpe

Nature takes over the ground on which once was laid the single line under Knutsford Rd. swing bridge. When the line was in existence, trains would have to wait for the bridge to be turned before they were able to pass through, sometimes for considerable periods whilst the bridgemaster waited for an expected ship. Platelayers had an even worse problem when carrying out maintenance work.

D. Thorpe

A view of the area of land between Acton Grange and Old Quay, Runcorn, which was sometimes referred to as 'the island'. In the centre right of the photograph is the cinder path where a few years earlier, the single line of the Manchester Ship Canal Railway was laid.

D. Thorpe

The sorry spectacle of scrap locomotives surrounded by oxy-acetylene cylinders and cut-up platework at Mode Wheel.

Courtesy H. Hudson

A line-up of 'Austerities', Kitsons and two of the large Hudswell Clarke saddle tank locomotives at Ellesmere Port, awaiting their ignominious end. Note the modified three tread footsteps on the nearest 'Austerity' locomotive.

Courtesy P. Warbey

some 300 yards east of Moore Lane. It was a pity that so much time and effort had gone into resiting the line, for eighteen months later it was closed down completely. Traffic to Wigg Works had finally come to an end after over 70 years. Laporte, at that time, was still using the railway with no plans to discontinue its use. Unfortunately, the volume of traffic from the plant was insufficient to make it an economic proposition from the Company's standpoint and in January 1970 the section was closed.

Joint BR-Manchester Ship Canal Working Party

In an attempt to end the problem of loss-sustaining sections, the Company, together with representatives of British Rail, set up a Joint Working Party as far back as 1966. Its brief was to study the effects of changes in pattern and volume of railway traffic, particularly as they affected shipment traffic. One of the tasks of the working party was to forecast the potential traffic passing to and from the Port of Manchester and the various industrial estates. This included an assessment of the operating and maintenance costs of the Manchester Ship Canal Railway network including the working over Trafford Park lines. An interim report was submitted in March 1967 which mainly dealt with costs, quotations of rates, the selling of services offered, and the setting up of a joint statistical body. Before the final report was published, some of the results of the findings of the working party had been implemented, the closing of Weaste Junction to all general traffic being one of the more visual signs. In June 1970 the final report was published and the recommendations resulted in the closure of the dock railways, the withdrawal of services to most of the firms on the Trafford Park and Canal estates, including Barton dock Estate and Stanlow, and the scrapping of most of the Company's rolling stock. The report did not, in fact, recommend the closure of the dock railways, it actually stated that the Manchester Dock traffic paid its way, for BR that is. The idea of cutting docks' operating costs by concentrating on handling selected berths only was, however, not recommended as it was thought it would lead to unacceptable limitations on the berthing of ships. With the majority of the track being bedded in concrete, there would have been little saving in maintenance costs even if selected berths were used.

Trafford Park

In view of the diminishing receipts from firms in Trafford Park and Barton Dock Estate, the Company issued a notice of termination of railway services which was to come into effect three months from the date of issue. If a firm could give assurances that it would be using its line for traffic at an acceptable level, the notice of termination would be withdrawn. By the end of 1970 when the notice came into effect, 52 firms had declined to give such assurances, and the following year a further 50 firms followed suit, which resulted in the services being withdrawn from most of the firms formerly using the railway.

Many of the traders were not too pleased at losing the docks and railway facilities, and one in particular, Reif & Son Ltd., was concerned enough to take the Manchester Ship Canal Company to task with a bid to retain the railway service, arguing that it was these very facilities which had attracted the firm to Barton Dock Estate in the beginning. Normally, the firm's timber, which originated in Yugoslavia, was taken overside direct on to railway wagons and delivered to the firms siding. If a complete cargo was being moved, it would involve several train loads. Unfortunately for Reif & Son, the firm was unsuccessful and the railway service was terminated. Ten years later the firm was still importing considerable quantites of hardwood and all of it is brought in by road, but the firm still considers the earlier system to be more useful.

With the closure of the dock railways and loss of most of the industrial traffic, the Company's rolling stock was of no further use, apart from a few wagons and coaches retained for work purposes. Consequently, during 1973, most of the rolling stock was gathered together for disposal. For the best part of eleven months the lines of empty wagons squealed and rumbled towards Salford, where they were stored awaiting despatch to the scrapyard. When the last wagon trundled along the main route it seemed to signal the end of an era; the silent empty marshalling yards, the sidings overgrown with weeds, the through route coming alive for just a few moments each day as a locomotive passed along on its way to a duty at either Weaste or Trafford Park. Stanlow still had its trains of tanks and, at that time, Partington was still dealing with the coal and soda ash traffic, but the best days were gone. Coal stages and water tanks were either in a dilapidated state or had been demolished, and of the ten railway weighbridges formerly in use, only two remained, one at Partington Coaling Basin, the other at 29 Stores which served Irwell Park Wharf.

During 1970 the Company decided to sell a certain number of locomotives as the fleet, at that time totalling 40 diesel units, was obviously too large for the Railway Department's greatly reduced needs. The first two, DH15 and DH16 were sold towards the end of the year to Bowater Ltd., for use at the Ellesmere Port plant. Surprisingly, the new owners retained the original Manchester Ship Canal numbers, although the livery was changed from blue to red. Between 1970 and 1976 the remaining two steam locomotives, together with 15 diesel locomotives, were sold, which left 25 locomotives in the fleet.

As a direct result of the docks railway closure the staff was reduced from 230 to a mere 146, which was drastic enough, but when one considers that the reduction in staff from the beginning of dieselization onwards was from a figure of over 550 down to that figure of 146, then it will be easy to imagine the feelings of those members of staff who were affected by all that had happened. As one ex-railwayman said, 'we knew

The end of the steam era with empty, weed-surrounded, coal stages and redundant water-columns. The coal stage once served the engines at Weaste Junction whilst the water-column stands at Pomona Docks.

D. Thorpe

Locomotive No. 90, looking very grimy but still impressive, working on 31st October 1960.

M. D. Lister

At 7.45p.m. on a summer evening, locomotive No. 54 *Odessa* takes a mixed train through Trafford Park. The date is 6th June 1962 but the photograph could very well be captioned the 'Autumn of Steam'. The rather neglected look of the engine says it all.

I. Higginbotham

A view of the single line, with Barton Bridge faintly visible in the far right of the photograph. The timber yard is that of Sheppard Bros. Ltd., and in the left foreground will be seen the points which originally controlled entry to the timber yard.

D. Thorpe

Locomotive No. 59 brings in a train of empties at Irwell Park Minerals Wharf. As the wagons are filled, a second engine moves the train along until the whole train is loaded. A fresh train of empties is then shunted into position. This is known as 'setting the quay'.

Courtesy M S C Co.

things were not so good, but somehow it seemed that the work would come back to us, as it had done in the past'. Even one of the more senior members of staff described himself shaken by the turn of events which had almost erased a once seemingly everlasting railway.

The National Union of Railwaymen and the Transport & General Workers' Union representatives were involved in the planned scaling down operation from the beginning. It was realized that the Company could not conjure up some magic formula to maintain the existing staff levels in the face of falling receipts. Naturally, the National Union of Railwaymen was particularly concerned for it represented the majority of the men affected. At the Salford branch alone it had over 500 members, therefore, it simply worked to obtain the best deal possible.

For a period of six years after 1972 the situation remained more or less static. Railway activity was concentrated in six areas and apart from Partington and Eastham, played no part in the movement of ship cargoes. The remaining sections were only concerned with traffic to and from various industrial premises. Partington also handled a certain amount of non-shipping traffic in the form of tanks for British Tar Products. Since 1973, when the Manchester Ship Canal laid new sidings for the firm, British Tar Products had been receiving oil rail terminal block trains of fine chemicals, the firm having entered the business of bulk liquid storage. For the National Coal Board coal traffic, the Manchester Ship Canal were using coal tip No. 3 and No. 5, No. 7 tip being used exclusively for soda ash. At Eastham, Pak-Tank (Unitank), which again was a bulk liquid storage firm, was taking trains of caustic soda liquor from I. C. I., Runcorn. Also on the Ellesmere Port-Eastham Section, Bowater Ltd., was taking 6 trains of coal each week which amounted to almost 4,000 tons weekly.

At Stanlow the Shell Company was sending out trains of bitumen, lubricating oils and paraffin whilst Associated Octel provided a substantial amount of traffic in chlorine and ethylene dibromide. In the Manchester area there were nine firms being served viz; Kellogg (GB) and Containerbase Ltd., on Barton Dock Estate, Esso, Petrofina, Rank Hovis, British Steel Corporation (Taylor Bros.), Corn Products Company and Ciba-Geigy (only one tank in a while) and at Weaste, Lancashire Tar Distillers.

Closure of Weaste-Partington Line

During 1976 the mineral wharf at Irwell Park was closed and the following year it was decided that, in view of the absence of revenue earning traffic between Weaste and Partington, maintenance on the line would cease, as costs attributed to the section had reached an unjustifiable level. Consequently, on Saturday 10th December 1977, the Railway Superintendent, Mr B. Valentine, issued a memorandum ordering the discontinuation of its use for traffic engines. Activity on the line was confined to engineer's traffic, usually the exchange of locomotives for repair or overhaul, the last being D3 from Partington to Mode Wheel shops and the replacement locomotive, D7 back to Partington on Wednesday 21st December 1977. Towards the end of July 1978 the through route was finally severed at Irwell Park prior to the lifting of the section. This was done in stages between 1978 and 1980. It is all the more regrettable to see this particular line disappear for, as was stated earlier, it was one of the most difficult sections of the Manchester Ship Canal system to com-

Although lacking the excitement and dash of the earlier steam-hauled trains, there is, nevertheless, something stirring about watching a diesel hauling its train up a stiff incline without stalling. This scene was witnessed on 13th April 1982, as Rolls-Royce Sentinel No. 3004 crawled up 'The Bank' at Stanlow with a train of bitumen tanks.

Miss E. M. Bannister

Railway swing bridge No. 2 being used for the last time. Diesel-electric locomotive No. 4001 moves slowly across on its way to Barton Dock Estate. Aboard were Driver Harold Rossitter and Shunter Bob Hamblett.

David Thornley

Partington Coaling Basin (north side) and a deserted south side. The line shown in the picture is the through route which formerly led to Latchford.

Courtesy M S C Co.

Partington (North) weighbridge which, until December 1977, housed the single line control operator for the line between Irwell Park Eccles and Latchford.

D. Thorpe

plete. With its removal the route and track length was reduced to 13 and about 40 miles respectively.

In the early part of 1979 the country was swamped in a wave of strikes, two particularly protracted ones being those of the lorry drivers and the road tanker drivers. The immediate effect on the Manchester Ship Canal Railway was the interruption of the Containerbase trains, but there was also the long term effect as industry generally took the brunt of the strikes. However, there was a small amount of traffic gained, albeit temporarily, when Proctor & Gamble Ltd. started moving its materials and products by rail. In fact the firm's transport manager attempted to set up a regular service using large covered hoppers, but unfortunately the rail system of Trafford Park was not strong enough to bear the weight involved and the scheme did not materialize. Further traffic was lost during August and September when the Kellogg factory on Barton Dock Estate was brought to a near standstill by striking employees.

1980 saw the closure of the Bowater plant at Ellesmere Port with the loss of 1,500 jobs for the town and, for the Manchester Ship Canal Railway, 196,000 tonnes per year of coal traffic. This left just three firms being served from Ellesmere Port West Junction; Pan-Ocean Terminals, Unitank, and at Ellesmere Port West sidings, Jake Ltd., rail tank cleaning services. The firm had, in 1979, established a depot on the site of the demolished Ellesmere Port locomotive shed, and were handling about 21 x 45 tonne bitumen tanks each week.

1980 also saw the Company's hopes for a revitalized Partington dashed when the NCB awarded Garston Docks the contract to handle its coal exports. This came as a body blow for the Manchester Ship Canal Company, whose engineers had prepared a detailed design of the proposed new coal handling system, and with the Company prepared to invest £1.25 million in the project it was generally felt that the coal traffic would be retained. To add to the troubles at Partington, the I. C. I. soda ash traffic, about 60,000 tonnes per annum, ended at the end of 1980 leaving only British Tar Products traffic, about 80-90,000 tonnes per annum, to be dealt with.

Partington Coaling Basin (north) was scheduled to close at the end of March 1981, but due to delays in the completion of the Garston facility, the NCB persuaded the Manchester Ship Canal several times to postpone the closure and to continue handling coal traffic. Partington eventually closed on Friday 3rd July 1981.

During the intervening months discussions had taken place between British Tar Products the Manchester Ship Canal and BR concerning the moving of British Tar Products traffic. Finally, a reluctant BR agreed to use its locomotive to do the job and a trial run was made over the Manchester Ship Canal lines down from Glazebrook Junction. After certain work was done on the lines at the insistence of BR, the block trains were brought in, the first being delivered on Thursday 9th July 1981.

Another loss of traffic, albeit a very small one, was the discontinuation of the Esso traffic to and from the firm's depot on Trafford Wharf. The service ceased on 31st August 1981. Then came the second most regrettable incident of 1981, after Partington, when the lines between Mode Wheel and Railway Swing Bridge No. 2 were lifted. On Friday 31st October at 3.45p.m., locomotive 4001 crossed the swing bridge on its way to Barton Dock Estate. This was the last engine to do so before the track was lifted. The bridge was afterwards turned into the clear position for shipping and the 86 year old link was finally broken. It was a grey, wet afternoon, the weather being ideally suited to the mournful occasion.

However, there were some gains made that year. On the site of the former Bowater plant at Ellesmere Port, two new firms were established, Ellesmere Stevedoring & Forwarding Company Ltd., taking over the wharf section, whilst the works area was taken over by Consolidated Bathurst, a Canadian-based firm. The new firm commenced to modernize the plant, the intention being to begin producing paper by the middle months of 1983. Unlike Bowater which made its own pulp at the plant, Consolidated Bathurst plan to buy in the pulp, thus reducing the power needed for the various processes (the plant produces its own electricity); therefore if the coal used is to be brought in by rail as before, the quantity will be at a reduced level.

The Manchester Ship Canal suffered a serious set-back in the last month of 1981 when a 100 tonne tank was derailed, lying on its side, thus completely blocking the line to Eastham. Between 10th December and 23rd December, all rail traffic to and from Unitank and Pan-Ocean Terminals was halted. Apart from the loss of revenue, the Company also lost out in having to pay BR for the use of a heavy breakdown crane.

Although certain specific instances of reduced or lost traffic have been discussed, these were but a few of the problems facing the Manchester Ship Canal Railway. However, it is not a simple matter to point to any particular set of circumstances and state that these were the reasons for the decline in the use of the railway; there are so many factors. Some of the more obvious factors are common to most railways; rising costs being one, but by far the most telling was the gains made by the road haulage body. Road transport had been making good ground for several years, and with the growth of the motorway network and the introduction of the container system, road transport was obviously in a strong position. This was further enhanced by the increase in cross-channel roll-on roll-off vehicle ferries. At last the ideal of a door-to-door service was possible, something that railways can rarely achieve.

Shipping using the Canal had also declined due, in the main, to the increased size of vessels. However, even before size became a problem, ships were not calling at Manchester in the same numbers as before, and the area itself was experiencing difficulties with factories closing down or moving to other areas, which all added up to lost tonnage for the Manchester Ship Canal. With the closure of Irlam Steelworks and the general decline of the steel industry, iron ore and specialist ores were no longer being imported. In Lancashire, 'King Cotton' no longer reigned; man-made fibres, if not wearing the crown, had largely replaced them.

Another bulk traffic loss for the railway was sulphur which, at one time, was handled in considerable quantites in rock form. Originally brought in from transatlantic sources in fairly large ships, the sulphur would be unloaded at Irwell Park Mineral Wharf, a process which would take about two or three days. Apart from a relatively small quantity which was sent round to Wigg Works via Latchford and Acton Grange, the bulk of the cargo, sometimes as much as 10,000 tons, was taken to Courtaulds, Trafford Park. Once at the plant the sulphur would be unloaded by grab crane and finished off by men with shovels. Obviously this was a slow and laborious task, and with the introduction of tipper lorries which speeded up the operation, the railway lost the traffic.

By that time the sulphur was coming in from Poland and this allowed the use of smaller ships, therefore, the cargoes were smaller which made the use of lorries more attractive. However, even that service ended after the sulphur began to be imported in a liquid form at high temperatures. The Manchester Ship Canal built a high temperature handling facility at Runcorn and this left Irwell Park Mineral Wharf without any traffic, and in 1977 it was closed to shipping altogether.

Cement was another commodity which was moved by the railway in considerable quantities, and was brought in by BR direct from the manufacturers to Manchester. It was taken to Turners Asbestos Company in Trafford Park and was then combined with asbestos to make building materials. With the reduction in the production of these items the quantity of cement was likewise reduced and in 1976 the railway lost the traffic.

As with other privately-owned railways connected to BR, the Manchester Ship Canal Railway felt the effects of certain events on its larger counterpart, or as one railwayman put it, 'when BR sneezes the Canal catches a cold'. It is certainly true that an extension of delivery time caused through industrial action by BR employees, or by BR policy, would have an adverse effect on the attitude adopted by users of the Company's railway. However, not all BR policy changes were to the disadvantage of the Manchester Ship Canal Railway and its users; the introduction of block trains, the larger capacity wagons and the Freightliner trains are but a few which have encouraged certain firms to continue using the railway system.

Referring once again to the problem of industrial action, the Company also was constantly affected. The dreary spectacle of idle docks brought to a standstill by strike action had become all too frequent, although on only two occasions were members of railway departments involved in a stoppage; the first during the 1926 General Strike, the second on 3rd October 1962 when locomotive drivers and firemen staged a one day stoppage. It is a fact that labour relations between the Company and its railway departments has always been fairly harmonious. The Mechanical Engineers had only one strike, but this was a fairly protracted one which occurred between 27th April 1983 and 31st May 1983.

As the 1970s advanced, industrial trouble of a more serious nature manifested itself as world trade slumped. By the beginning of the 1980s the trade recession was bad enough to be compared, by some, with that of the 1930s depression. Many industrial concerns, including large and well-established firms, were forced to close, which added to the growing unemployment. Obviously shipping companies were in the same position, many of their ships being laid up and staff reduced.

Apart from the closure of Partington Coaling Basin, the loss of shipping to the Canal had little effect on the railway and indeed, it was one part of the Company's operation which paid its way. On an even brighter note, both shipping and railway tonnage showed an increase in the closing weeks of 1981. However, there did not, in the early part of 1982, appear much room for optimism, the start of the year being marred by a series of strikes by ASLEF members of BR. Obviously this affected every section of the Manchester Ship Canal Railway at a time when extra traffic was available.

In fact, 1982 again showed a welcome rise in the shipping tonnage totals, an increase of around half a million tonnes. However, the railway tonnage totals showed a drop of around 7,000 tonnes. Even so, taking into account the closure of Partington, and the losses due to industrial action by people unconnected with the Company, the figures were not too bad, some sections actually showing an increase.

The tonnage figures for both shipping and railway were heartening, but the situation as regards the Manchester and Salford group of docks was decidedly not. Most of the shipping using the Canal was, by 1982, being handled at Ellesmere Port and Runcorn. There were only three regular callers to the docks at Salford. With the announcement by Manchester Liners, on 10th June 1983, of the ending of its Mediterranean service to Manchester, the large container terminal at North 9 Dock was virtually made redundant. A second line, the Atlant Line, operated by The Anglo-Yugoslav Shipping Company, also ceased its Manchester service, which left only about five or six ships being handled each month.

The possiblility of transferring the remaining up-Canal shipping to the Eccles-Barton section seems more than ever likely. This would leave the whole of the docks area open for industrial development as was already happening around 6 Dock. For many people, especially the unemployed, any development was welcome, but to others there was a sadness in seeing the docks reduced to such a state.

The Manchester Ship Canal's fortunes reflect those of the Manchester area; in the mid-1950s, long before the Japanese-Far East competition had really started to be taken seriously, and long before the EEC agreements had upset established trading patterns, production capacity in the Manchester area had started to decline. The area's machine tool, diesel engine manufacturing, and locomotive building industries are but a few of those which have declined or disappeared completely.

Possibly what the area, and the country, needs at this point are more Daniel Adamsons to generate a new commercial and public enthusiasm, both of which are needed if it is to witness, once again, an era of industrial and social well being.

Looking to the future, there are changes which could be beneficial to the Canal and its railway; one being a decision by the Shell Company to build smaller ships which may lead to the building, by other companies, of vessels which will be able to negotiate the Canal. It is not an impossiblity that a dock railway system could be reintroduced at a future date. There is certainly a possibility that the Manchester Ship Canal will, in 1994, celebrate its centenary and still be operating its unique railway system.

Chapter Five — Trafford Park

Just as the Manchester Ship Canal had been born of the desperation of a declining city, and a visionary father in the shape of Daniel Adamson, Trafford Park, the world's first industrial estate, was born of the opportunities presented by the Canal and fathered by another visionary, Mr Marshall Stevens. Without doubt, if Manchester and the surrounding towns had not been in such a precarious situation, the Canal would not have been constructed and Trafford Park would possibly still be the home of the de Trafford family. Conversely, without Trafford Park, the Manchester Ship Canal Company would have had a very difficult time of it in the early years, and would probably have not become as busy as it did.

Before the coming of the Manchester Ship Canal, Trafford Park was truly a park consisting of 1,200 acres of meadows, farms and woodlands where deer roamed. In the pleasant rural setting, with its beautiful lake, was Trafford Hall, a mansion which compared favourably with many of the dignified houses in the county.

Pre-History 1066-1896

The Hall, at the time of the Canal construction, was the home of Sir Humphrey de Trafford whose family had lived there and owned the land since before the time of William the Conqueror, in 1066. Earlier, in 1016, an ancestor of the family, Ralph Trafford, had been given a small portion of land by King Canute in return for his military exploits on the King's behalf. Later, after the Norman conquest, King William gave all the lands, between the River Mersey and the River Ribble, to Roger of Poictiers. Part of the land was, in turn, granted to Baron de Macy who then confirmed the de Trafford of the day his holdings of land at Stretford, which included that which was to become Trafford Park.

On the northern side of the estate, a natural boundary was formed by the River Irwell, and after 1761, when the Duke of Bridgewater opened his historic barge canal, the estate's western and south-eastern boundries were formed by the Worsley and Manchester branches of the Bridgewater Canal. Thus the 'Park' became virtually an island.

On completion of the Manchester Ship Canal, the younger Sir Humphrey de Trafford decided to sell the estate, and in 1896 Mr E. T. Hooley, a financier, purchased it with the intention of developing part of the area as public pleasure gardens. He was also considering the building of several villas, probably of a more exclusive type, and another scheme which was considered was the introduction of a new racecourse.

The Ship Canal had been open for just over two years by that time, but the struggle to break the grip of Liverpool, on trade and shipping, was proving extremely difficult. Trade and family ties between businessmen in Liverpool and Manchester were still very strong, therefore, the answer was to create or attract industries to Manchester which had no such ties. The forming of Manchester's own shipping line, Manchester Liners, was still 2 years away and the Canal was only being used occasionally by large ships.

To attract new industries to the immediate area was, with the release of Trafford Park for development, then a real possibility. Lying alongside the Ship Canal, having the Bridgewater Canal on the other boundaries, and with a railway network surrounding it, the transport advantages were obviously excellent.

Mr Marshall Stevens who, at that time was the Manchester Ship Canal Company's first General Manager, was convinced that this was the solution to the problem which was affecting the people of Manchester as well as the Company and, therefore, he approached Mr Hooley with the proposal to develop the 'Park' as an industrial estate. His proposals were taken up, and on 17th August 1896 the new company was launched. So convinced was Marshall Stevens on the importance of the venture that he resigned his post with the Manchester Ship Canal and took the position of Managing Director of Trafford Park Estates Ltd.

The plan was to lay out the area into suitable plots which would be served by a network of roads and railways, and for this purpose the services of Mr Robert Edis, the respected municipal planner, were employed. However, it was soon realized that to attempt the task of forward planning in such a manner was not feasible.

Most of the difficulties arose because of the size and diversity of the premises. Some factories were small and some were able to be built in neat rectangular sites, but most of the factories differed in area and sometimes in their odd-shaped site requirements. Obviously, the only way to tackle the layout of the estate was to deal with each factory site individually. The laying out of roads was a comparatively easy task, but with the railway it was not so simple. A minimum radius curve of 200ft. had to be adhered to and this parameter alone caused problems in the layout of sites, some of which lay at very odd angles in relation to other sites. There was a certain amount of track laid, which, it is recorded, was connected to the Manchester Ship Canal system by a temporary connection as early as 1898.

On 1st November of that year an agreement was reached between the Estates Company and the Manchester Ship Canal which gave the Estates Company the right to a permanent railway connection with the Manchester Ship Canal system. This in effect gave Trafford Park Estates an access to the railways of the main line companies, at that time the LNWR, L&YR and Cheshire Lines Committee. From the agreement the Manchester Ship Canal obtained free running rights over the estates lines and were also allowed to collect a fee of 6d per ton on any loaded wagon carrying goods for traders in the 'Park' which had paid Ship Canal tolls. In brief, goods paying tolls for shipment on the Ship Canal itself. The concessionary rates were extremely cheap, and coupled with the twice daily delivery and collection, made the Trafford Park Docks railway system very attractive to the firms on the estate. This, of course, was the whole idea, but as years went by the rate of inflation, albeit a low one for many years, gradually whittled away at the profit margin, making the Trafford Park Docks traffic less attractive to the Manchester Ship Canal. It was to be over 50 years before the rate was altered and even then the rate was still very low. In 1899 the Estates Company obtained an Act enabling them to construct and operate railways in Trafford Park. This was done under the West Manchester Light Railway Order 1899.

A new Act, called the Trafford Park Act of 1904, made the Trafford Park Company the operating company for the Trafford Park Estates railway. The new Act enabled them to extend the railway where and when it was required, without the need to petition Parliament each time an extension was planned.

Early days in Trafford Park at a time when there were still wooded areas and cattle were still being grazed. Although, from the photograph, it is not possible to discern much of the detail in the background, on the original print it was possible to see animals grazing by the Estates' boundary wall, whilst beyond the wall could be seen a line of ships, including a graceful paddle steamer.

Trafford Park Estates Ltd.

Part of the developing estate looking westward towards Barton, circa 1915. In the far distance (left) are the oil storage tanks of General Petroleum Co, Ltd., Homelight Oil Co. Ltd., and General Oil Storage Co. Ltd. In the far distance, (right) is the bulky mass of Togo Mill which lies on the Eccles side of the Canal. Note the lengths of line ready for laying and also the early car, possibly a Model T Ford.

Trafford Park Estates Ltd.

The offices of Trafford Park Estates Ltd., in 1981. Apart from the types of vehicles and the shrubbery in front of the building, little has changed since they were first built.

D. Thorpe

The older part of the Estate with factories built very close to each other. It was a restrictive environment with firms having no room for expansion. Even so, many of the firms have stayed for more than half a century.

D. Thorpe

Although there is no doubt that *Sheriff* was a Trafford Park Co. engine, its origins and age are unknown. By the style of the wheels and the chimney, it could be an Avonside Engine Co. product and, by the damaged front end, it had been involved in a fairly heavy collision.

Trafford Park Estates Ltd

The Hunslet locomotive, No. 15 *New York* is pictured with its train on Trafford Park Rd. outside the works of Henry Royce, Electrical Engineer, more widely known through his partnership with Charles Rolls and their Rolls-Royce cars.

Trafford Park Estates Ltd.

The Trafford Park Company laid lines along both drives which originally served the 'Hall' from the entrances at Trafford Road and Barton. An earlier tramroad had been conveyed to the West Manchester Light Railway by the Estates Company in 1903, subject to a payment of £300 and an annual charge of 1d for each additional yard of track put down.

A service of passenger and goods trams was introduced in conjunction with the British Gas Traction Company, but the gas driven trams were only used until 1910 when they were replaced by the more powerful electric trams. Apparently the gas trams, when fully loaded, sometimes failed to negotiate the bridge over the Bridgewater Canal at Barton, and only managed to get over with the aid of the more energetic passengers putting a shoulder to the struggling vehicle!

In 1905 a junction was formed with the Midland Railway via the Cheshire Lines Committee siding at Stretford. To enter the Cheshire Lines Committee system originally, it had been necessary to pass into the Manchester Ship Canal system through 'A' Junction then through the Manchester Ship Canal-Cheshire Lines Committee connection at Cornbrook. The new connection was made via Bridgewater Junction, which has been previously mentioned. The bridge carrying the lines over the Bridgewater Canal was usually referred to as the 'Manchester Light Railway bridge' by the Divisional Engineers' staff of the Bridgewater Department. With the opening of Bridgewater Junction, the positions, with regard to railway connections for the Manchester Ship Canal and the Trafford Park Company to the lines of the Cheshire Lines Committee, were reversed, for Cornbrook Junction was then closed. This meant that the Manchester Ship Canal traffic now had to reach the lines of the Cheshire Lines Committee through Trafford Park. However, the right to run over the Trafford Park Company's lines without payment meant that, apart from the extra distance travelled, it was no change for the Manchester Ship Canal, that is except that the steep gradient up to the viaduct was, from that time onwards, avoided.

The following year an agreement, which rescinded a previous 1902 Agreement, between the Cheshire Lines Committee, Trafford Park Company, Manchester Ship Canal, GNR, GCR and MR was signed on 28th March 1906. It also gave the Cheshire Lines Committee and, in effect, the GNR, GCR and MR, the right, in perpetuity, to exercise running powers over the Trafford Park Company lines. This was done by the Manchester Ship Canal acting as their agent in Trafford Park.

In the agreement the Cheshire Lines Committee agreed to pay the Trafford Park Company the annual sum of £2,200 and tolls of 1d, 2d and 3d per ton, according to the class of traffic carried, for traffic in the area known as the Green Area. This sum was referred to as the Commuted Sum. The Green Area referred to was an area shown on the estate plan coloured green. A second area was shown coloured pink which was, of course, the Pink Area.

By 1916 the track length had reached 13¼ miles and ten years later the figure had more than doubled. Although by 1946, with a total of 33 miles, the layout of the system had more or less reached its final stage of development, the odd additions here and there added a further two miles to that figure by the late 1960s.

Laying the track was the responsibility of the Works Department of Trafford Park Estates Limited. An internal agreement was reached on 20th May 1977 concerning the construction of the railway system. All previous work had been done by the Works Department acting as a separate organization within the Estates Company.

For the construction and maintenance of the permanent way the Works Department employed about 10 to 12 platelayers, a further 15 to 20 men being employed as crossing keepers. There were also several roving crossing keepers who could be sent to any section as the need arose. The crossings were denoted by the letters A, B, C and G. As part of their equipment, the keepers were supplied with flares for use in foggy weather.

In 1919 a set of regulations covering the railway system was brought into force, as follows:

1. Speed restricted to 10m.p.h.
2. No more than 30 wagons per train except in case of shortage.
3. System of engine whistles.
4. Traffic to be worked by shortest route.

At a later date the maximum speed was raised to 15m.p.h. and the number of wagons allowed on one train was increased to 50.

During the first five years of the estates existence, the number and size of firms moving in must have been very satisfactory to the infant company. By the year 1902 there were 31 industrial premises, a power-station, 2 banks and a housing development. The largest factory and employer of labour in Trafford Park's history, British Westinghouse Electrical & Manufacturing Company Limited, was already established and had just commenced production. Their work-force was, at that time, over 3,000 and still increasing.

Although the Westinghouse Company had bought the site (Plot A) as early as 10th June 1899, they were not the first to settle in the Park. There were three firms which had moved in

Each site was offered complete with railway connection and this was laid by the Estates' subsidiary company, The Trafford Park Co. In the photograph a new factory is being built and the railway is being laid. Note the steam roller working on the site, seen in the middle distance.
Trafford Park Estates Ltd.

A white-coated crossing keeper allows MSC Kitson No. 74 to cross the cart road. Apart from the crossing keepers stationed on major road rail crossings, Trafford Park Estates Ltd. employed 'roving' crossing keepers. These men were sent to any point which temporarily needed supervision.

R. S. Greenwood

An MSC Hudswell Clarke long tank locomotive hauls a mixed train out of 'B' sidings, ('B Junction' to the Manchester Ship Canal men). Another engine of the same class stands on the far side of the train. The shipping seen in the far distance (right) is in the graving docks of the Manchester Dry Docks Co.

Trafford Park Estates Ltd.

A load of cotton on one of the TPE wagons. This particular wagon was very similar to the contractors bufferless rolling stock used on the Canal construction.

Trafford Park Estates Ltd.

Trafford Park Estates bogie wagon No. 27 with a rather forward thinking firm's motif painted on the sides. The combined 'T' and 'P' for *Trafford Park* was painted on all engines as well as on rolling stock. Note the restraining chains and plates on both bogies.
Trafford Park Estates Ltd.

(Above) Apart from the manufacturing and servicing firms which were established on the Estates, there were several warehouses built, including those of the LNWR and the L&YR. The photograph shows the entrance to the L&YR warehouse viewed from across Railway Road (now Trafford Park Road).
Trafford Park Estates Ltd.

A works train, headed by *Sir William Bailey*, (Hudswell Clarke 0-4-0 saddle tank, (Works No. 841) built in 1908, stands alongside Railway Road (later Trafford Park Rd.) near the steelworks of Taylor Bros. Ltd. The John Biesty lorry is a Leyland vehicle.
Trafford Park Estates Ltd.

during the first year but, unfortunately, the records do not show who the earliest tenants were. Certainly Morrison, Ingram & Company, the sanitary engineers, W. T. Glover, the cable manufacturers, and Lancashire Dynamo & Motor Company, were all in the estate quite early, but it is possible that some of the smaller and lesser known companies were established at an earlier date. It has been suggested that the earliest factory was that of N. Kilvert & Son, the edible fats processor, but they, in fact, did not arrive until 1903.

In 1902 the number of firms' premises served by the railway was at least 26, certain of the other premises being on sites which lay on the Chester Road side of Trafford Park which, due to the position of the lines of the Cheshire Lines Committee and the Bridgewater Canal, were not in a position to be connected to the Trafford Park railway system.

As well as factories, there were numerous sites taken for warehousing facilities. Apart from the warehousing companies, there were properties belonging to the LNWR, L&YR and the Cheshire Lines Committee and also the Port of Manchester Warehousing Limited. The last-named company was an associated company of the Trafford Park Estates. Trafford Park Estates had first entered the warehousing business in 1906 when they built four large warehouses, which had a total capacity of 6,120,000 cubic feet. They were built close to several other warehouses which belonged to the Manchester Ship Canal, who subsequently took a lease on the Trafford Park Estate buildings. Due to the demands of war, the Estates Company increased its warehousing capacity by 3,675,000 cubic feet in 1915, and by a further 4,575,000 cubic feet the following year. The Port of Manchester Warehousing Limited was formed to handle all warehouses belonging to Trafford Park Estates Limited.

The 1922 Agreement of 17th February, between the Trafford Park Company, the Manchester Ship Canal, the Cheshire Lines Committee and the LNWR, reaffirmed the position whereby all traffic was carried by the Manchester Ship Canal. At the same time the payment made by the Cheshire Lines Committee to the Trafford Park Company, known as the commuted sum, was increased to £2,500 whilst the tolls were increased from 1d, 2d and 3d per ton by an agreed percentage.

Several firms employed locomotives of their own for internal shunting. British Westinghouse are believed to have been the earliest private user of engines in the Park, their 0-4-0 Manning Wardle being active around the works in 1902. Eventually they were employing three steam locomotives before diesels were brought in. Turners Asbestos employed at least two engines, these being named *Turnall* and *Asbestos*. Barton Power-Station also employed two locomotives, both being 0-4-0 saddle tanks built by the Avonside Engine Company.

Alongside 'A' Junction lay a set of sidings which was originally used by the Southern Cotton Oil Company. The firm employed a locomotive which bore the name *Scobrit* which is, of course, a partial acronym of the firm's name. At a later date the company, and the locomotive, became part of Kraft Foods Limited. Other firms operating locomotives of their own were Esso, Procter & Gamble, Redpath-Dorman Long and C.P.C.

As the years advanced the number of firms establishing themselves in the Park grew, and so did the diversity of products. Electrical equipment, including motors, dynamos, control gear, and water driven turbines, foodstuffs, constructional steelwork, timber, motor cars, railway coaches and thousands of other items. Henry Royce, of Rolls-Royce fame, established his main works in Trafford Park, in 1902, to manufacture, amongst other things, electric cranes, and the American Car & Carriage Company took premises in 1904, and from this firm came the carriages for the London Underground Railways.

A large site was taken up in 1907 by Taylor Brothers & Company Limited, makers of railway wheels and axles. A fairly extensive internal railway system was put down which was worked by their own fleet of locomotives. In 1911 the Ford Motor Company established their British assembly plant in the Park. Parts were brought in from the United States and delivered to Manchester via the Ship Canal, then moved to the plant by the Manchester Ship Canal Railway.

A glance at the list of names of companies with premises in Trafford Park was akin to looking at a *Who's Who* of international industries. There was the engineering firm of Sir W. H. Bailey & Company, the head of the firm being a

Hudswell Clarke engine (Works No. 1250) *Lord Ashburton*. The Trafford Park Estates' locomotives were to have a relatively short stay in 'The Park', all being sold about 1921, after the Manchester Ship Canal Co. had objected to their use as traffic engines to and from various firms' premises.

Trafford Park Estates Ltd.

Trafford Park Estates locomotive *Sir Joseph Lyons*, looking very new, and its crew. This engine and 'fellow' engine, *Lord Ashburton* were built in 1916 by Hudswell Clarke & Co. and were of the same type as those owned by the Manchester Ship Canal Co. Works numbers were 1249 and 1250, respectively.

Trafford Park Estates Ltd.

One of the well-known names and also one of the old established firms in Trafford Park, 'Brooke Bond OXO' with its pleasant frontage. It is the growing practice amongst companies in the Estate to lay lawned areas, and to plant shrubs and flowers around their premises.

D. Thorpe.

An aerial view of Trafford Park with the various works sites belonging to AEI highlighted. Main Works, as it is usually referred to, is seen in the lower centre of the picture.

Courtesy Switchgear Testing Company Ltd.

A comparatively latecomer to the Estate was Ingersoll-Rand, yet another internationally known concern.

D. Thorpe

Some firms chose sites alongside the Canal which gave them the facility of a private wharf. To the right of this picture is the plant and wharf of Corn Products Company Ltd. (CPC), whilst in the middle distance is the pipe bridge and factory of Proctor & Gamble Ltd. (formerly Thomas Hedley Ltd., soapmakers).

Courtesy J. Steggles

Timber merchants have always been part of the scene in Trafford Park. These are the yards of Parker Rosser (formerly Frank Rosser & Co.) and James Rowlinson & Sons (M/cr).

D. Thorpe

director of Trafford Park Estates Limited, International Combustion, Dorman Long (a firm of structural steel engineers renowned for such notable structures as the Sydney Harbour Bridge, the Storstrom Bridge in Denmark and many others), and Redpath Brown, another structural steel engineering concern. Turners Asbestos, makers of construction materials and other products, Carborundum and Ingersol Rand, both American companies, Esso, Burmah Oil, Shell and Manchester Oil Refineries for oil products, Brooke Bond Tea Limited, Lyons Tea and Kraft Foods Limited - these were just a few of the manufacturers whose names were well-known to most people. The head of Lyons Tea, Sir Joseph Lyons, was also a director of the Estates Company.

In 1931 a soap making firm, Thomas Hedley & Company of Newcastle, built a large factory alongside the Ship Canal, the site being ideal for the discharging of oil from the ships direct into their plant via over-road pipes. In July 1962 the firm became Proctor & Gamble Ltd.

The Swiss Chemical concern, the Geigy Company, inventors of the insecticide DDT, established a factory near the Barton end of the Park. Later the premises were occupied by the Ashburton Chemical Works Ltd., which is still part of the Geigy Group.

Timber importers and merchants have been a part of the Trafford Park scene since the earliest days, Illngworth Ingham & Company and Manchester Timber Importers being amongst the first to arrive on the estate. Other firms are Frank Rosser & Company Ltd., John Ashworth & Company, G. Mason & Company, Parkers (Ancoats) Ltd., and James Rowlinson & Son.

With the outbreak of World War I in 1939, a heavy burden was placed on the factories and work people of Trafford Park. The estate, for the second time in its history, became a huge centre for the production of essential materials and weapons of war. The workforce multiplied, large numbers of women being brought in to replace the men who had gone to the armed forces. From an estimated 50,000, the numbers of people employed rose to 75,000.

The Park was producing aircraft and aero engines, guns, mobile power-stations, radio equipment, drugs, medicines and a variety of less obvious but, none the less, important products with which to aid the war effort.

I. C. I. set up a plant for producing penicillin, the Ford Motor Company established a new factory for the production of Merlin aero engines, the number of engines coming from the plant being 34,000. Amongst the many things produced at Metro-Vickers were Lancaster bombers, these being built at a new factory in Mosley Road.

Obviously, the vital role played by Trafford Park was known to Germany, with the result that for several nights on a run, the estate was subjected to bombing attacks. This was the infamous Christmas Blitz of 1940. There was a certain amount of damage but not enough to stop production; many of the workers at factories which had sustained damage, including those which had lost roofs, simply worked on, in some cases, in the open air.

With the ending of the war in 1918, the inevitable drop in output caused the tonnage figures to fall, and although they recovered and showed a steady increase over the following 20 years, the amount of traffic carried on Trafford Park Railways appeared as a declining percentage of the total Manchester Ship Canal figures. The period 1939-45 however, showed a surprising similarity to that of World War I when the percentage of Trafford Park Railway tonnage against the Manchester Ship Canal Railway tonnage was in the region of 36. During the war, the total tonnage carried over the lines of

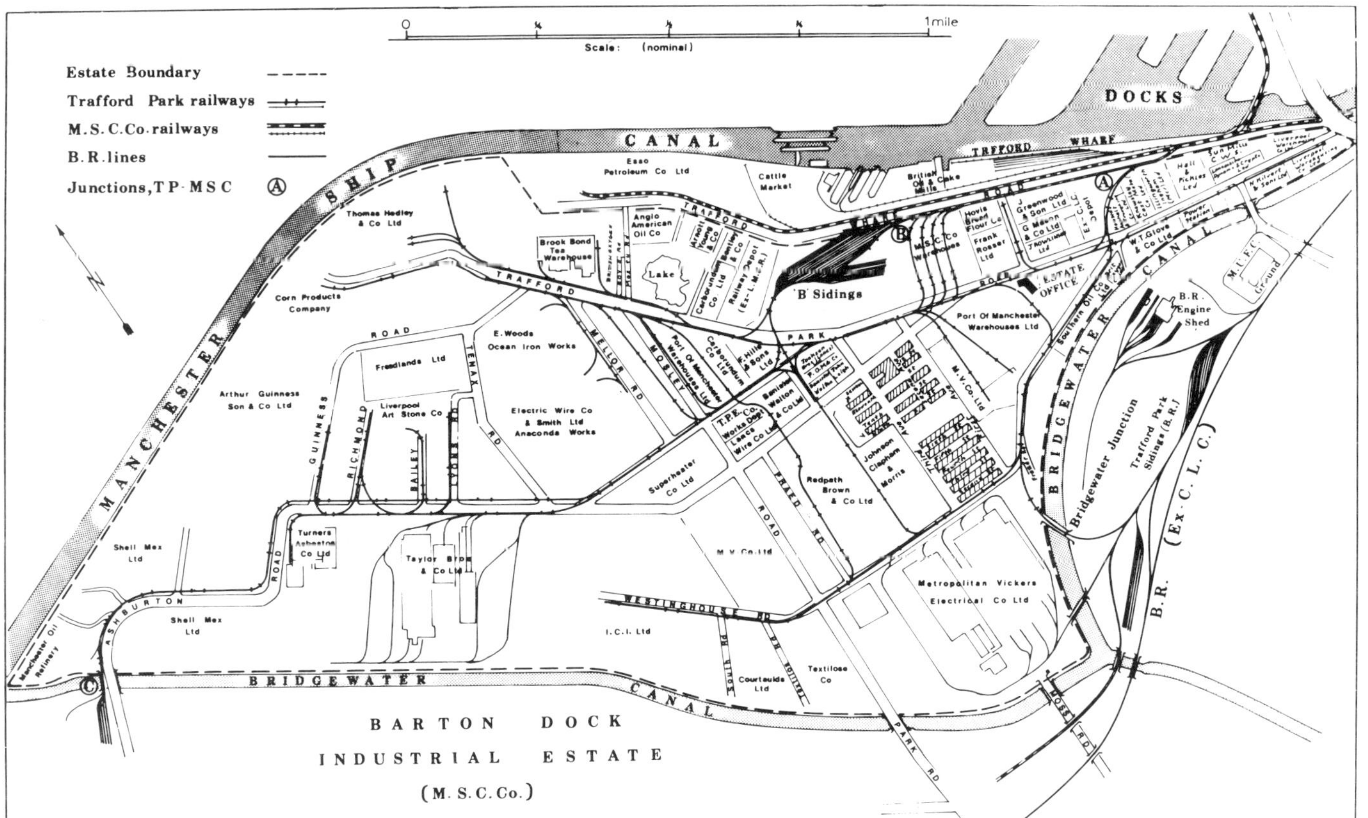

the Park was 13,000,000, whilst the Manchester Ship Canal carried over 40,000,000 tons, the Park traffic averaging 32.5 per cent of the Manchester Ship Canal traffic. In 1931 it was estimated that the tonnage moved over the Parks railway system was 3 per cent of the country's total railway tonnage. Coal for the use of Trafford Park firms alone reached 300,000 tons annually and that was in a period when the railway was taking less of the available traffic as each year passed.

Decline in Use of Railway

Road transport was gradually taking a larger share, and years later with the opening of the M62 motorway and, later the M63, on completion of the Barton High Level Bridge in 1960, the advantage of road transport became even greater. In the following years, the motorway system grew to a point where today most areas of Britain are in direct connection by motorway with Trafford Park.

Both the Trafford Park Company and the Manchester Ship Canal were suffering large losses on their respective railway operations, and the closing of the dock railways in 1972, followed by the scrapping of most of the Manchester Ship Canal rolling stock soon after, changed the whole pattern of traffic in the Park. This, of course, followed the publication of the BR-Manchester Ship Canal Joint Working Party Report of July 1970.

That year the Trafford Park tonnage percentage was down to 16.55 and two years later the figure had dropped to 12.9. Over the next six years the tonnage totals of both the Manchester Ship Canal and Trafford Park steadily decreased with only slight fluctuations and, by 1979, when the Manchester Ship Canal Railway actually carried an increased tonnage, Trafford Park's contribution was down to 6.99 per cent.

Joint Working Party TPE Company, MSC Company and BR

Obviously, the effect of dwindling traffic and receipts was felt by all parties concerned, consequently a joint working party was set up, in 1969, comprising members from the Trafford Park Company, the Manchester Ship Canal Company and BR. Their brief was to examine the whole operation and, in particular, the expected future traffic pattern over the following ten years.

The report of the Joint Working Party was published in October 1972. It is an extremely comprehensive survey of the situation existing at that time, the principle conclusions and recommendations being as follows:

Traffic:

Traffic to and from the principle traders on the estate makes a modest contribution to BR system costs after allowing for payments to both the Manchester Ship Canal Company and the Trafford Park Company, both now and in the future. These traders are expected to retain their importance as users of the estate railway, at least during the next ten years. (Paras. 24-30, 33-34).

Track:

The fall in traffic levels, coupled with the Manchester Ship Canal Company's decision to withdraw rail service from the docks, gives the opportunity to rationalize the rail network, which has remained almost unchanged for over fifty years. (Paras. 35-36). A survey of the track, plus the Manchester Ship Canal Company's estimate of a new operating pattern, showed that the route between Bridgewater Junction and 'B' Sidings (and the sidings themselves) could be abandoned, the system remaining being sufficient to serve the main traders on the estate and provide adequate connections to the Manchester Ship Canal system. Several other lengths of track will be abandoned reducing the track mileage from 26¾ to 6⅓ (Paras. 37-43).

(a) the smaller system envisaged
(b) the absence of renewals, the system (as brought up to standard by mid-1973 needing planned maintenance only during the next ten years.

The Working Party recommends that the tolls payable to the Trafford Park Company be based on the Company's estimated expenditure in future years. We also recommend that the tolls be on a tonnage basis and that the annual fixed sum payment be discontinued. The new arrangements should be introduced from 1st January 1973 (Paras. 44-53). The tolls recommended should be subject to a price variation clause, on the lines of the one now in force between BR and the Manchester Ship Canal Company (Para. 63). The increase in the Manchester Ship Canal Company's operating costs should be met by increasing the terminal allowance paid to the Company by BR (Para. 52).

Contact between the Parties:

The Working Party recommends that the regular contact already existing between the parties, be extended to cover all matters which could impinge on railway working (Para. 66).

Legal Matters:

The Working Party have felt it to be outside their province to express any views on the legal position if their recommendations are accepted (Para. 68).

One of the recommendations concerning the track length had, by the end of 1979, almost been carried through to the letter, the reduced track length being just over 8 miles which is almost down to the 6⅓ miles suggested. The withdrawal of services to those firms giving only spasmodic and small quantity traffic, was effected after being assured that the firms in question were able to make alternative arrangements for the transport of their former rail borne traffic, at no substantial increase in costs.

'B' Junction and Sidings have been dispensed with, all movement between the Manchester Ship Canal system and that of Trafford Park being effected through 'A' Junction, at the time of writing, consisting of Manchester Ship Canal locomotives travelling to and from their duties in Trafford Park and Barton Dock Estate. As a point of interest, the original Manchester Ship Canal Acts relating to Trafford Park have never been officially cancelled.

Changes in Trafford Park Industries

Although there have been many changes in the type of industries which have, at some time, been located in Trafford Park, the vast majority of firms have been concerned with manufacturing, particularly in the engineering field. However, within a space of twenty five years, since 1950, the nature of the work carried out on the estate has changed dramatically. Although a good number of the original firms are still in existence, some with new names, many of the larger manufacturing concerns have either drastically reduced their workforce, set up factories elsewhere, or closed down altogether.

The largest employer of labour, AEI-GEC, has reduced its labour force by about 70 per cent, BSC (Taylor Bros.) has also reduced its labour force and its diversity of products due to rationalization within the British Steel Corporation

Carborundum Co. Ltd., makers of abrasives, is another famous name, the factory being established in Trafford Park during 1913.

D. Thorpe

The large traffic island near the western entrance to Trafford Park against a backdrop formed by the laboratories and plant of the Ciba-Geigy Co.

D. Thorpe

Many of the premises being built now are meant for the smaller businesses and the group of buildings shown in the photograph are very representative of the size.

D. Thorpe

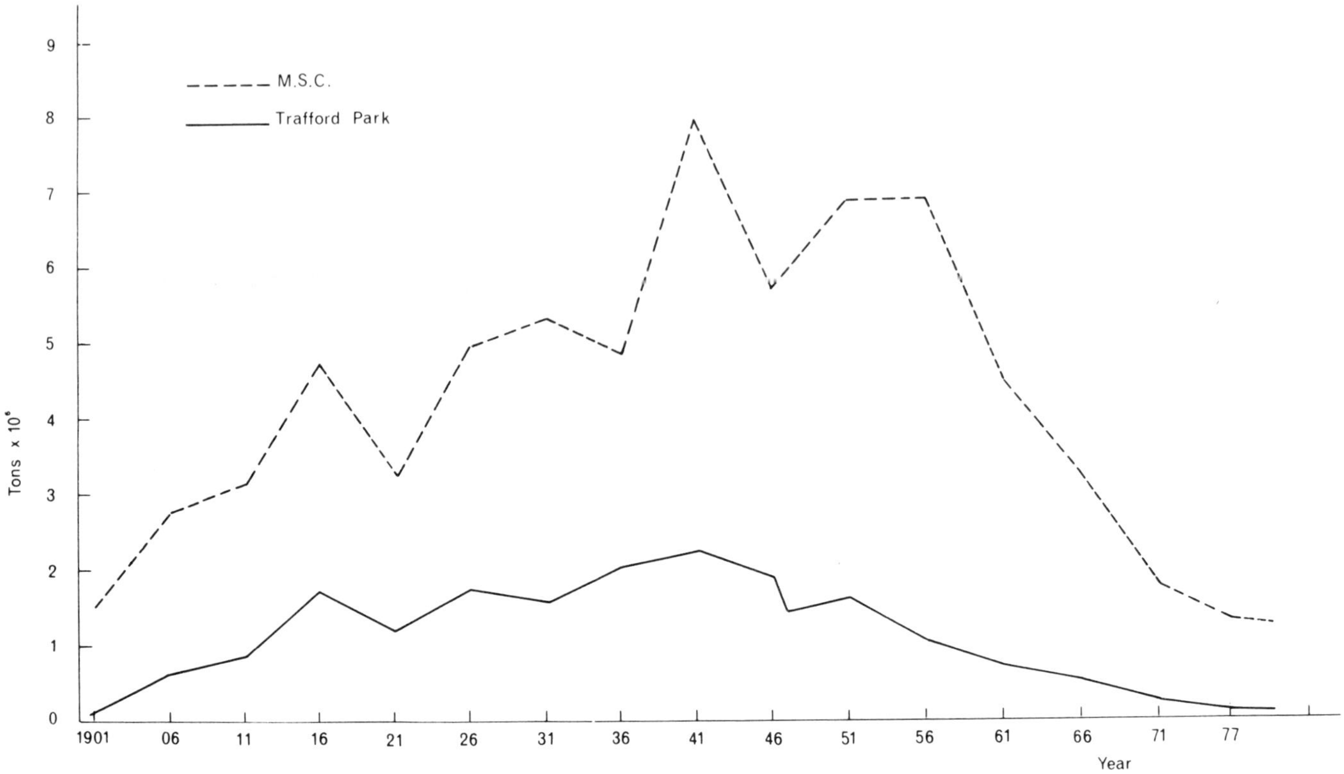

national structure. Another large plant which was in the process of closing down during 1980, was TAC (Turners Asbestos Company).

In the place of these industrial giants has stepped a number of smaller firms, but a good proportion of these are concerned with either service or distribution activities.

Outwardly the Park is quieter today; it is also cleaner, thanks to the Clean Air Act. Along some roads are grassed-over sections where, at one time, the lines lay. Much of the track which is visible is unsightly even to a railwayman; dirty ballast supporting rusty ribbons of steel no longer being polished through usage. The local authority, who maintain the roads, are not very enthusiastic about these railway tracks, as they mar the landscaping which has been carried out under the plan to make the area more attractive.

Business continues to flourish however, the visible signs being new buildings amongst the old, and also the volume of commercial road traffic. Successive trade recessions, Government monetary policies, and the increasing competition from foreign products, have all hindered the industrial growth of the area and yet it continues to chug along. Trafford Park is a name which, to this day, is synonymous with industry and still retains its unique advantages - ideal transport facilities, a large pool of skilled labour on its doorstep, and there is still space for development.

Chapter Six Staff and Operations

Staffing the Early Railway Departments

On completion of the Canal the task of recruiting workers had not been as easy as it was at first supposed. It is true, there was a large pool of labour to be drawn from, for the unemployed were scrambling to obtain jobs with the infant company. Unfortunately, the type of labour was, at that time, considered unsuitable for many of the jobs to be filled, therefore, dock workers and others with previous experience were brought in from other ports to form a basic labour force which required no training, and which would be able to train others who came after. In contrast, the staffing of the engineering posts and the railway departments was made up of a nucleus of men who had previously been involved in the same sort of work during the construction period. In addition, other experienced men, many of them former employees of the main line railway companies, were enrolled after 1894.

As the Manchester Ship Canal Railway developed and expanded, the number of persons wholly occupied with railway matters increased to a point where they represented the largest group serving any one section of the Company's activities. Apart from those only partially employed on railway work, including certain clerical and workshop staff, the number wholly employed on railway work was, at its peak, at least 784.

The Railway Department Staff including, it is thought, the Railway Superintendent and his assistant and also the Superintending Inspectors.

B. Valentine

A group of Manchester Ship Canal Co. Inspectors. Note the ornate uniform and badges.

B. Valentine

There were, and still are, three departments whose divisions of responsibility cover the railway. The Railway Department is, of course, solely concerned with railway operations, but both of the other departments, the Resident Engineer's Department (Civil Engineers) and the Mechanical and Electrical Engineer's Department are, however, involved in a great deal more than just the work of the railway. To the Resident Engineer's Department falls the job of canal bank maintenance, masonry work on locks, bridges and quays, and also the maintenance of all Manchester Ship Canal buildings. The Mechanical & Electrical Engineer's Department deals with every mechanical and electrical item on the Ship Canal. Railway weighbridges are part of the Department's responsibilities and are maintained by the Scales Department at the workshop which was situated in 6 shed, 8 dock. Signals, however, are the responsibility of the Resident Engineer's Department, the signal fitter working from the shop at Salford Yard.

There were several other members of staff who worked on, or for, the railway, and who were not regarded as railway staff, certain of them not even coming within the spheres of the Railway Superintendent, the Chief Mechanical & Electrical Engineer or the Chief Engineer. Two such departments were the Railway Rates Section and the Railway Settlements Section. Both departments worked with the Railway Department and, in fact, shared the same floor in the Dock Office, but acted as separate bodies. The Railway Settlements Section came under the Accounts Department.

Traffic Department

Most of the men employed by the Railway Department began their working lives as messenger boys. By this it meant

A mixed group of Railway Department Staff, including Chief Superintending Inspectors, Inspectors, foremen, shunters and cabin lads.

B. Valentine

A train, headed by locomotive No. 90, rumbles across No. 2 railway swing bridge at Salford on its way to 'B Junction'.

M S C Co.

that a boy would be attached to one of the, usually, wooden offices which housed the Inspectors etc. Those on the clerical side started as office boys. For many years the practice was for the messenger boys to sit a form of examination to assess whether or not they were suitable for a clerical position. Cabin boys, however, had no form of assessment on the manual side, for after a boy had served some time as an assistant labeller, and a further period as a number taker, he would then be 'streamed' into one of two paths. He could work either as a labeller followed by the job of wagon follower, or he could take the job of shunter. A shunter could advance through eleven grades to the position of Railway Superintendent, but for the wagon follower there was only one step up and that was the position of stock inspector. Actually, there was only ever one position of stock inspector, and it was not always filled.

After some years as a shunter, a man could possibly become acting foreman shunter, after which he could climb the promotion ladder to foreman shunter and on to Inspector, Grade 4. A further three grades would make him Chief Inspector. The Chief Inspector at Salford was always deemed to be senior to Chief Inspectors at other stations. To reach the position of Railway Superintendent, a Chief Inspector would have to be promoted through the grades of Assistant Superintending Inspector, Superintending Inspector and Assistant Railway Superintendent. Actually, this climb to the top only happened once and that was in the case of the last Superintendent, the late Mr B. Valentine; Inspectors and above tended, previously, to be drawn from the administration section.

Under the direction of the Railway Superintendent were several non-railway staff, including the supervisors at Partington, Runcorn and Ellesmere Port. In later years the latter three positions were dispensed with, since, as related earlier, Runcorn had lost its railway system and the other two sections came under the direct control of the Railway Superintendent.

Of the 361 manual staff which covered twenty seven different occupations, the largest number in any one occupation was the shunters, of which there were 137 plus 19 foreman shunters. On the other end of the scale, numerically, was one weighman, one lampman and one stockman, all at Salford, whilst at Ellesmere Port there was one checker. Apart from the jobs mentioned, the department employed 66 sheeters, 7 sheet repair men, 15 foreman loaders, 3 other weighmen, 3 traffic regulators and a host of others; telephone attendants, checkers, crossing keepers, timekeepers, signalmen and tipmen. Two groups of workers not mentioned above are the number takers and labellers. During World War II female labour was employed; 5 on labelling at Salford, 5 on number taking at Stanlow and Ellesmere Port. After the war only the female number takers were retained; by the 1970s only two lady number takers were being employed, both at Ellesmere Port.

Staff at Runcorn

During the years when the railway was in operation at Runcorn, a staff of 14 was employed on railway work. This figure included the permanent locomotive driver and the driver of the shunting horse, but not the horse! The situation

Ellesmere Port yards on 11th August 1955. Locomotive No. 89 pauses a while as the 'lady numbertaker' books each wagon. During World War II, females were employed in various capacities on the Manchester Ship Canal, and at Ellesmere Port there are still female numbertakers employed to this day.

R. H. Appleton

Loading coal on No. 5 tip in June 1977. On the lower line, a loaded wagon waits its turn to be emptied into the ship's hold. The wagon on the top line is moving off the tip, aided by gravity.

D. Thorpe

A winchman manoeuvres a wagon on No. 5 coal tip. Wagons are hauled into position from the feed line, turned on the turntable, seen in the centre, to present the end door of the wagon to the chute. They are then moved into position on the hydraulic lift. Wagons can also be received from the line feeding in from the sides of the tip.

D. Thorpe

A train of empties makes its way up the gradient from Partington South to Glazebrook Junction. Hudswell Clarke diesel-mechanical locomotive No. D14, with its Gardner 8L3 power unit, makes light work of the climb and, although the drivers are all ex-steam men, they readily admit that it takes a lot to stop this particular class of locomotive.

D. Thorpe

with regard to footplate staff was a bit irregular at Runcorn. The shunter acted as fireman, whilst the spare locomotive was driven by one of the engineer's staff who was, in fact, a crane driver.

The administration for the railway has always been based at Salford, which is only to be expected as the vast majority of traffic, both shipping and railway was, until the mid-1960s, handled at this, the terminal end of the Port of Manchester. Originally the Railway Department's Offices were sited with the early Dock Offices on the Old Trafford side of the Canal near to Trafford Road swing bridge. At a later date the department moved to offices in Salford Yard. At the peak of railway activity, in 1943, the Railway Department employed 174 clerical staff and 361 manual workers. Not included in these figures were the Railway Superintendent and his senior staff.

Apart from the normal clerical work connected with traffic, such as the administration, demurrage and standard charges, releasing traffic, advising and invoicing, there was a considerable amount of work concerning the requirements of private traders on such items as ordering empty wagons, sheets and ropes, etc. Having such a large outside staff entailed a fairly comprehensive record system covering such items as staff uniforms, shunting poles, sprags and shunters' lamps plus, of course, all the items mentioned previously which included sheets, ropes, etc.

The number of wagon sheets ran into thousands and, as can be imagined with hundreds of wagons being loaded and unloaded every 24 hours, the number of sheets which were damaged in the process added up to a considerable amount of work for the repair staff. The sheet repair shop was situated in Salford Yard, not far from Main Gate. Later, in the 1950s, the department was moved to its final home in the Pomona Docks area. With the disposal of the rolling stock, in 1973, the sheet repair shed was closed down.

Mechanical Engineers Department

Under the wing of the Mechanical Engineer's Department are the locomotive and workshop staff. For those intent on becoming a locomotive driver, the period between starting as a boy and reaching the position of driver was a lot longer than that taken by men in other jobs. A youth would work as a 'bar lad'; that is one who replaces firebars, then for probably one year as a cleaner, after which a further eight years would be served as a fireman. Before being given the job of driver, the fireman first had to pass a test set by the Locomotive Superintendent. Drivers were in two categories; ordinary driver and merit man. The latter title was selfexplanatory and was normally awarded to drivers of long service, but on occasions other drivers were made merit men, when they were judged to be of above average capability. Being a merit driver put extra money in a man's pocket and bestowed upon the man the additional perk of always having the same engine. There have been many merit engines over the years, the numbers changing as locomotives were brought up to a better standard through overhaul or rebuild. To have a locomotive on a permanent basis was a fine idea, for in each case the engine would benefit in attention to both mechanical parts and cleanliness. In the earlier years all locomotives were cleaned and polished to a very high degree, no oily rags or paraffin swabs being used on the paintwork, only soft soap and warm water. Regrettably, due to the demands on men and machines during World War II, regulations concerning the cleaning of engines were relaxed and, as a result, locomotives were given a wipe over with oily rags, which turned the smart colourful livery into a uniform black. Even so, some engines were kept very clean and the merit engines were amongst the best.

The war also brought other changes, one of these being the length of time served by a fireman before he became a driver. Due to the shortage of staff, a fireman could then take the examination for driver after only six years firing. From the ranks of the drivers came the Locomotive Foremen and the Locomotive Superintendent. The Locomotive Superintendent and his staff were based at Mode Wheel Workshops.

Running shed staff, such as boiler washing out men, sand dryers and coal heavers were on regular employment in these capacities. Although exact figures are not available for the Locomotive Department's staff, the total, including the

Sanding up at Mode Wheel locomotive shed. This operation had not changed since the railway was formed. Even after the sand was dispensed from the hopper, boxes were still filled in this way.
Courtesy H. Hudson

A yard scene at Mode Wheel Workshops, circa 1960, which could be rephrased as 'the morning after the day before'. Here the yard gang are clearing the fire droppings left by the weekend's locomotives. A Hudswell Clarke short tank is on duty as yard engine with the yard foreman, Joe Coxon, on the footplate. In the left of the picture is the Boiler Shop with the pay coach standing in front. Slightly towards the centre is the front of the stores and main office block, partly obscured by the front of the Fitting Machine Shop. Partly obscuring that shop is the Wagon Shop extension with both doors open. The high building behind the locomotive is the original Wagon Shop building with the Joiners' Shop to the rear.

Courtesy J. Steggles

Mode Wheel locomotive shed in its original form. The three roads were known as Cemetery Road, Middle Road and Sandhole Road, the latter because it ran alongside the sand drying furnaces. Originally envisaged as an engine shed to cater for 35 locomotives, it is doubtful, even with the smaller engines of the time, if it would accommodate more than 30 engines.

J. Steggles, Courtesy W. Clayton

Mode Wheel locomotive shed in the 1960s with a new roof, complete with a single longitudinal smoke vent and, with the entrance wall removed, it was a lighter and less smokey place to work in. However, during the winter, conditions were much worse with the wide open front.

Courtesy H. Hudson

Mode Wheel Shed today, or what is left of it. To the left of the picture is Sandhole Road with the 'White House' (workshop) at the end.

Courtesy B. Valentine

Clearing the quay at No. 9 Dock. Locomotive No. 32 *Gothenburg* works its last days before retirement.

M S C Co.

Superintendent, his foremen, the drivers, firemen and all other personnel, was in excess of 290.

Charlie Chambers

No account of the Manchester Ship Canal Railway would be complete without mention of the late Charlie Chambers, locomotive driver. His was a typical story of the early days. He came from a poor family and started work in 1891 fatting wagons on the Canal site. His age at that time was 12, and he obtained employment by producing his older brother's birth certificate. As time passed he graduated to driving steam cranes, locomotives and, in his own words, 'going anywhere where there's a bit of steam and fire, and that's where I will probably end up'.

Much of the credit for the details presented in this account of contract locomotives and sheds has to go to Charlie Chambers, whose memories of his early hard life on the building of the Canal were still so clear until the end of his life in 1977.

Workshop Staff

Mechanical Engineer's workshop staff employed full time on locomotives, wagons, etc. numbered about 65, and were mostly tradesmen and apprentices involved in either locomotive fitting, boiler making or wagon building. Others employed on railway work, but only for part of their working day, also included some tradesmen such as machine shop staff, plumbers, painters and blacksmiths.

Resident Engineer's Department Staff

The Resident Engineer's railway staff was usually kept at a strength of about 55 but, during 1948, the figure had risen to about 60. These figures include platelayers, gangers and point oilers.

Railway Operations

A good way to illustrate the workings of the organization as a whole is to describe some typical railway operations. The first example concerns an imported cargo of phosphate. Information would be received at the offices of the department, to the effect that 9,000 tons of phosphate would be due at Manchester some time in the near future. An estimated time of arrival would be ascertained, then the operation started. Instructions would be issued for the gathering together of as many 16 ton steel mineral wagons as possible. Trafford Park was a good source from which to draw, as the firms there 'made' hundreds of empties every week. This is where the importance of the number taker's work is demonstrated, for their job was to keep a tally of each wagon whilst it was on

The end of the shift at Mode Wheel. Merit driver Frank Shawcross (left) leaves his engine to the ministrations of the shed staff. The Shawcross family were one of those with ties going back to the construction period, Frank's father being one of the 'navvy crackers', as the contractor's engine drivers were called.

Courtesy H. Hudson

Working on the quays was a dangerous business with little room between the lines of wagons. Once the train was loaded or unloaded, as the case may be, the train was taken out and a fresh one put in. This was usually done during the hours of 12.30-1.30p.m. and 5.30-6.30p.m. If the resetting of the quay was asked for between those times, it was referred to as an 'intermediate' shunt.

M S C Co.

A scene at Mode Wheel Locks, looking towards Manchester, as the tugs *Arrow* (out of sight) and *Sovereign* manoeuvre the *MV Brasov* into the lock. To the left of the picture is the railway track referred to as the 'Old Route'.

J. Steggles

the Manchester Ship Canal or Trafford Park system. Instead of sending the wagons 'home', that is, returning them to their owners, they would be marshalled on to 2 Grid where they would be swept clean by the docker/sweepers. Wagons would then be tared on the weighbridge after which they would be stored in sidings around Irwell Park Wharf. About 550 wagons would be needed and the task of obtaining and storing them would be made even more difficult if several bulk cargo carriers were due about the same time. As the loading proceeded, the wagons would have to be moved along until each one had been filled, at which time the road would be cleared by one locomotive, then reset by another. Each loaded wagon would be taken over the weighbridge, labelled and then taken to one of the main line junctions to be put into the exchange sidings. In some instances the bulk cargo would be ore of a special grade, possibly chromium. For ores of this nature, the hoppers to be used would be swept out and then hosed down. This operation was usually carried out on the lines known as the 'old route', which was on No. 3 Grid. If wooden wagons were used, the planking would sometimes have to be caulked to prevent the ore escaping.

When ordering wagons up for special destinations, the wagons were chosen from the rolling stock of the railway company which served that particular area. During the period of World War II, all wagons were pooled, therefore any railway company's wagon could be chosen to go anywhere. Out of this situation the name 'poolers' emerged, a name which was given to box wagons and persists to this day. Wagons belonging to private firms are called 'owners'.

In dealing with 'exports', that is, inward traffic to the docks, a different process altogether was used. Taking one of the Manchester Liners as an example would be appropriate, for these ships always 'loaded out' at Manchester after discharging their cargoes. A firm, possibly from the Midlands, wishing to send a load of steel tubes to Canada, and perhaps a second manufacturer in Middlesbrough with a consignment of steel bound for the same destination, would contact BR who would, in turn, contact the Manchester Ship Canal Railway Department and 'offer' the traffic. The department would contact the shipper or cargo agent with the details. The shipper or cargo agent would know how his ship was progressing and would usually accept the 'offered' traffic so that it would arrive as near as possible to the time when it was needed. The Manchester Ship Canal Railway would then 'release' the traffic, the business of 'releasing' being done to ensure that the right party was responsible for demurrage. When it was on hand, all parties would be advised and the shipper could then order it in, very often by wagon number, so as to adhere to correct ship's stowage. Liaison was maintained between the shipper and the quayside delivery clerk who actually ordered in wagons through a person called a 'wagon runner'.

When rounding up wagons to move a ship's cargo, the choice of wagons could become very much like a juggler's act. In the case of a bulk cargo such as the shipload of phosphate previously described, the choice of wagons was a relatively

Locomotive No. 35 and a train of coal at No. 1 Grid. It is interesting to note the excellent brick facing of the warehouses. This was done to appease the earlier racecourse owners, at a time when the racecourse occupied the ground on which the immediate two sets of lines were laid.

M S C Co.

Ex-WD locomotive No. 87 heads a train of coal, on a beautiful sunny afternoon, as it passes Irlam Locks on 30th July 1957. The line branching off to the left of the photograph leads to the lines of the ex-Cheshire Lines Committee to Irlam Station which was used exclusively by the CWS since 1896. Where the train curves, a set of wagons stand in the exotically named Magadi Loop.

Alex Appleton

straightforward task. For the removal of a mixed cargo, which usually split up for several different destinations, the selecting of rolling stock could become extremely complicated due to the differing rules on the part of the railway companies. For instance, Sowerby Bridge, HM Dockyard Devonport, Birmingham, and Middlesbrough were just a few of the places which would not accept wagons with a longer than ordinary wheelbase. This was also the rule for many of the London stations. A weight restriction of 12 tons was imposed on vans for Darlington and 20 tons on vehicles for Chorley. Other stations could not take vans at all. At Sheffield City Station, a maximum of 21ft. length was imposed on all wagons. In the years up to 1939, wagons also had to belong either to the railway company whose metals they were to travel over, or to the private trader whose commodity was being carried.

Train Running

Whilst most of the locomotive duties were taken up with the tasks within sections such as those previously described, a substantial number of engines was involved in train running. From the junctions to the docks, from the Park to the docks, from Weaste to Irlam Steelworks the main route was busy throughout the twenty four hours. The longest runs were usually handled by the Partington men using the heavier classes of locomotive. Although train running was normally accomplished using a single locomotive, double-heading or bankers was not unusual, especially in certain local sections. Two such sections were the notorious 1 in 40 gradients known as the 'hump' between Salford Yard and Aubrey Street, and the 1 in 76 'bank' at Stanlow. Bridgewater Bank was another which often saw the use of bankers. There were quite a lot of other gradients to make life difficult for the footplate staff but most of these were tackled using a single locomotive. Acton Grange, Latchford and Partington all had gradients up to the main line junction and these had to be tackled with a full head of steam and plenty of determination. Another gradient, and one which looked deceptively easy, was that which lay on the through route between Daniel Adamson Road and Weaste Cemetery. On many occasions a locomotive would all but expire on the gradient whilst bringing a heavy train up from Weaste Junction.

One of the most exciting sights involving the use of double-headers, was that of a train leaving Salford Yard for Weaste. The crossing keeper at New Park Road cart road would alert the signalman at Aubrey Street who, when ready, would close the level crossing gates at Aubrey Street, then set the signals and points to give a clear line. The crossing keeper would then stop the traffic at New Park Road and signal the train on over the 'hump'. With a shriek from the whistle of the leading engine, the train would slowly get under way, gathering speed as it approached the climb. The roar of the exhausts blasting and echoing back from the Dock Office building, changing in note as the two beats came in and out of synchronization, was a sound to excite the imagination of any railway enthusiast. It was also a spectacle to see; those straining locomotives likened to heavy snorting horses thrusting with their muscular legs, the exciting mingled impressions of noise, of belching smoke and power. Once over the top, the tempo changed. With trains being unbraked, the job of slowing them down had to be most carefully judged. One driver who misjudged things later brought the locomotive into Mode Wheel with distinct flats on all six wheels.

To take a train up the 'bank' at Stanlow produced another similarly exciting scene but the method of banking, in this

A sunny day in May 1955 with locomotive No. 24 paused adjacent to the signal near Salford Shops. The engine, built in 1902 by Hudswell Clarke & Co., Works No. 642, started life as *Philadelphia*.

Courtesy Alex Appleton

A view on the lines of the ex-Cheshire Lines Committee at Partington (south side) with locomotive No. 19 *London*, a 'Jazzer', taking a train down to the tips on 26th March 1949.

C. H. Appleton

case, was always carried out using a locomotive at the rear end of the train. Each train taken up the bank, usually 20 to 25 loaded tanks if it was a Kitson, or 25 to 30 if it was an 'Austerity' or numbers 88 or 89, would be accompanied by a brake van, sometimes two if there were two trains of empties to be brought back down.

The duty at Weaste Junction, sorting wagons, was given the title 'Weaste bump' for obvious reasons. It was bump, bump, all day and all night long, and all regular shunting duties were referred to as 'bumps'. Thus a similar duty at Junction 'B' in Trafford Park was the "B' Junction bump', or, otherwise, the 'Galy bump'. An engine working this duty would be called upon to assist trains up the gradient leaving Junction 'B' in the direction of Bridgewater Junction. It is not clear why the term 'Galy' is used, but it was adopted in 1917 after the abortive British Military action at Gallipoli.

The duty known as the 'Barton bump' was not a sorting operation but involved the delivering of coal to Barton Power-Station and the reason for the 'bump' appendage was that, to reach the coal storage ground at the power-station, the train had to ascend a stiff gradient. Therefore, the operation had to be carried out in several stages with only six wagons being taken up at each stage.

These jobs were the normal lot of the footplatemen but occasionally they would enjoy a change, a train running duty, for instance, from Weaste to Partington. This run could be quite enjoyable especially for the driver and the shunter, but perhaps not so much for the fireman who had to work hard. A run to Partington would leave the more industrial scenery behind and travel alongside the Canal for some miles through semi-rural surroundings. Until the CWS Soap & Candle Works was reached, the surroundings were very pleasant; waving fields of oats and such like between Barton and Irlam, whilst on the opposite side of the Canal there was plenty of meadowland and trees to please the eye. Even the sewage works, on each side of the Canal, were not as offensive as some of the types of factory which were established elsewhere on the estate and in Trafford Park. One of the longest runs was probably from Partington to Latchford and again this could be very pleasant, for almost the whole distance was travelled through open country. Acton Grange to Wigg

The timber-built single road shed on Partington (south side). Outside stands a 'Jazzer' locomotive, No. 19.

B. W. Potts

On 1st November 1956, on the single line system between Rixton Junction and Latchford, 'Austerity' locomotive No. 87 moves towards Statham Tips where all Manchester Ship Canal refuse was dumped.

B. W. Potts

Works, Old Quay, was another section where the line ran through rural surroundings. Quite a lot of wildlife could be seen, rabbits, hares, foxes, and weasels. Certainly this environment was not in keeping with that usually associated with a port railway.

The locomotive crews seemed to enjoy their work whether it was in rural or industrial surroundings, but there were jobs which were not particularly welcome. Some of the men did not enjoy working trains of livestock, and most of them were dismayed when they were involved in dirty cargoes such as carbon black or dry slack coal. Even so, these jobs were taken in their stride except when the duty was also a midnight turn, at Christmas or New Year's Eve. These shifts were the cause of more grumbling than any of the difficult or dirty jobs, causing a great deal of disruption to the men's family lives and, ultimately, their stomachs.

At the Salford Section a job which was not particularly well liked was the 9 dock 1 Grid duty, usually referred to as the '9-1'. The turn would start with the job of sorting on 1 Grid after which a train would be made up and taken to 'B' Junction. After getting rid of that train, another one would be hooked on and delivered to Bridgewater Junction. A further train would be brought back to 'B' Junction, left there and the locomotive hooked on to a train usually referred to as the 'compo train'. This was a composite of various odds and ends which would be gathered en route for Weaste Junction. Loads would be picked up at Trafford Wharf, South 6, and Salford Yard. The 'compo train' would be left at Weaste and a final train brought back to 1 Grid where the shift ended.

A duty which was always a possibility was to be called upon to take firemen to the scene of a blaze if the need arose. Furthermore, an engine could be commandeered by the police at any time, even if it meant leaving the train. Actually, no instances have been unearthed of this ever happening but even today it is still in the Rule Book. In the event of such a thing occurring the driver is instructed to sound the whistle continuously whilst on his way to the fire. The vision of a locomotive bristling with helmeted figures, rushing along with the driver hanging on to the whistle chain, accompanied by the shrieking of the whistle, conjures up visions of the old-time film comedies of the Keystone Cops!

A cattle train leaves the Manchester Corporation lairages in the 1930s, headed by locomotive No. 40 *Buenos Ayres* in its original livery of green (by then much darker) edged in black and picked out in white. Transporting livestock was not a favourite job with many of the railwaymen, especially *if it meant putting in feedstuffs and water.*

M S C Co.

No. 1 Grid or New Barns, as it was referred to, where locomotive No. 14 *St. John* is seen bringing in a train from Trafford Park. From the lettering on the wagons, it will be seen that this was a pre-nationalization photograph with LMS and NE stock in evidence.

M S C Co.

Unloading timber from wagons in the lay-by of Morgan Wallworks Ltd. at Irlam, as a ship passes by on its way towards Barton Locks and Manchester.

Courtesy Morgan Wallworks Ltd.

Signals

As in most large dock railway systems, the layout of the lines at the various sections was quite complicated, and it took several years before a shunter or driver became really familiar with just one section. In addition to the complexitites of the track layout there was also the system of signals and engine whistle signals. To become familiar with the fixed signals on the Manchester Ship Canal Railway was not an enormous task, since there were only seventeen sets over the whole system. There were, however, eleven sets of signals affecting the Manchester Ship Canal locomotives, which were operated by outside authorities. At Ellesmere Port the passage between each section was, and still is, controlled by a semaphore type operated by BR (LNWR/GWR Joint) whilst the signal at Bowaters, again a semaphore type, was operated by that company's employees. At Acton Grange there were three signals, two being operated by Richard Evans & Company and one which controlled the single line between the Kemet Works at Astmoor Wharf and Wigg Works. The pair near the premises of Richard Evans & Company controlled the diamond crossing which incorporated part of that firm's railway line. At a later date, each of the three signals were operated by either the fireman or the shunter of the locomotive involved. All three signals were of the semaphore type.

Beyond Acton Grange there were no signals until the Barton Aqueduct set (Manchester Ship Canal) was reached. These two signals, both semaphores, were electrically operated by the swing bridge staff in the control tower.

The largest concentration of signals and indicator devices was at Salford, where there were eight semaphore types and one red/green light type. Six of these signals were operated from the signal box at Aubrey Street level crossing, which also controlled the level crossing gates and the red/green light signal on the crossing keeper's cabin at New Park crossing. This was the cart road opposite the main gate and close to the Dock Office. The set of lights were only in use when the crossing keeper was off duty which, normally, was overnight and at weekend from Saturday 12.30p.m until 7.30a.m. Monday. Both the northern and southern signals controlling trains over New Park Road were operated by the crossing keeper, whilst the signalman operated the six signals controlling trains in and out of 9 Dock loop to or from Weaste or Salford Yard.

Only one other Manchester Ship Canal signal box was in operation and this was situated on Trafford Wharf close to the railway swing bridge. This box controlled three semaphore signals also sited on Trafford Wharf, and also the basket ball signals at the bridge, which were intended for shipping movements between the main basin and Pomona Docks.

A Richard Evans & Co. (Haydock Collieries) engine, *Haydock* at the company's wharf, Acton Grange. Built in 1879 by Robert Stephenson & Co., (Works No. 2309), the engine sometimes worked at the colliery, its place being taken by one with the name *Magnet*. Apart from the export of coal, which was handled by a rail-borne crane, the wharf was used to accept imported pit props, as can be seen. The locomotive is now preserved in Penrhyn Castle Museum near Bangor, North Wales. The regular driver of the engines at Acton Grange was Freddie Biddel.

Courtesy F. Biddel

A charming picture, circa 1949, unlike that normally associated with docks railways, with Hunslet locomotive No. 84 and six-wheeled cashiers coach No. 1 on the single line to Irlam, after leaving Barton Locks. Lying at its berth near the locks is Manchester Corporation's sludge disposal vessel, *SS Mancunian*.

Courtesy A. Wadsworth

Waiting for the 'peg' at Barton yard. The engine, Hudswell Clarke locomotive No. 89, is taking a train to Partington and is about to enter the single line system.

B. W. Potts

A fireman's view of the Canal as locomotive No. 89 blasts its way along the single line system in the Barton Bridge to Barton Locks section.

B. W. Potts

Locomotive No. 78 under the newly-erected large capacity water tank at Partington North. The original standard water-column, seen in the background, was replaced because of the increased number of large engines stationed in the section.

B. W. Potts

A rather thought-provoking notice, warning drivers to be careful when approaching Cemetery Junction. This was the point where the lines from 2 Grid joined the west-bound through-route near Weaste Cemetery, a short distance before passing Mode Wheel Workshops.
Courtesy H. Hudson

Although the ball type signals were not intended for the railway system, the footplatemen used them as advanced warning of whether the bridge was being, or about to be, turned. All the signals at the swing bridge were interlocked with the bridge locking system, so that in the event of the signalman failing to set the signal to stop before the bridge was about to be turned, the action of unlocking the bridge would automatically operate the signals. This set cotrolled the movement of traffic between Salford Yard and Trafford Wharf. All the semaphore signals on the Manchester Ship Canal system were of the lower quadrant type.

A different type of signal altogether had been adopted by the Trafford Park Railway. They were of the 'banner' type, which was a white disc with a straight black band running across the centre from side to side. In the same way that the semaphore arm showed 'danger' when it was horizontal, so the black band would also indicate 'danger', and when the disc was revolved through 45 degrees, the black band, then in a diagonal position, would show 'clear'. These signals, of which there were three pairs, controlled the movement of rail traffic in the triangular section of line intersecting Westinghouse Road and connecting Bridgewater Junction and Junctions 'A' and 'B'. The passage of trains in or out of Bridgewater Junction is controlled by semaphore signals operated by BR, formerly Cheshire Lines Committee, signalman. They include the signal which controls the line leading to Barton Dock Estate.

As well as signalling devices, drivers had to learn the schedule of whistle signals. There were twenty five points at which whistle signals had to be sounded, and these were in addition to the normal warning whistle signals which was several short blasts in succession. Although there were some points where the signal was the same, the schedule was carefully arranged so that the points were in different sections. For instance, the signal for engines moving between Pomona and Trafford Wharf was simply two long blasts, and this was also used by engines entering or leaving Bowater Works, Ellesmere Port, Stanlow Junction or West Junction. Anywhere along the line between Wigg Works and Warrington Wharf, Latchford and Salford Yard, in the direction of Manchester, was the same as that from Irlam Junction to the Latchford line, two long and two crows.

From a driver's point of view, working in Trafford Park was a hazardous business, with the adventurous pedestrians and road users posing a constant challenge to their vigilance. There were several accidents over the years, some of them with tragic results. Usually the locomotive driver was completely blameless but, even so, many of them who had been involved in an accident was affected quite badly for some time after the event. In some sections the tracks threaded their way between high buildings creating an impression of working along the bottom of a canyon. On most of the straight sections the lines passed the entrance to numerous works and also the various side roads. Although the speed limits of 15m.p.h on straight sections and 10m.p.h on any curved section was laid down in the Rule Book for trains working in Trafford Park, many drivers would not travel at those speeds, being more cautious than others due to the conditions just described.

Although very few would have admitted it whilst they were working on the railway, most of those involved enjoyed their jobs. It was not unusual for men to spend the whole of their working lives with the Company. There was a feeling of being part of a vital operation and the staff had an attitude of involvement. On the locomotive side there were railwaymen's improvement classes which were attended by the men during their off duty time, and many of them bought books on the subject to improve their understanding of the machine under their control. There was an element of rivalry between the drivers, some of them being very jealous about the relative performances of their engines. A lot of the drivers would do

Unfortunately, few photographs are available which include locomotives of the railway companies working at the various junctions. However, one such photograph shows the ex-GCR 0-6-0 Class J10, No. 5183, with its crew, and others, standing in the sidings at Glazebrook Junction. Shunter Bostock stands against the tender. An interesting point to note is the windowed back plate fitted to the tender.
Courtesy Cliff Bostock

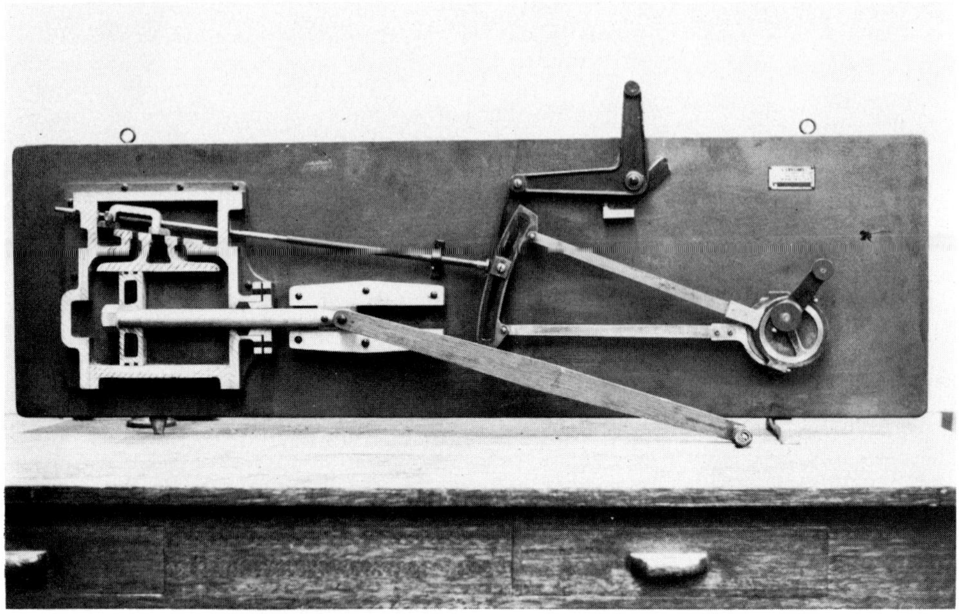

Two examples of teaching aids used at Mode Wheel, when instructing locomotive firemen the principles of valve gear and the importance of 'notching up' to conserve steam. The first photograph shows a proprietary-made item whilst that in the other photograph was made at Mode Wheel in the early years.

D. Thorpe

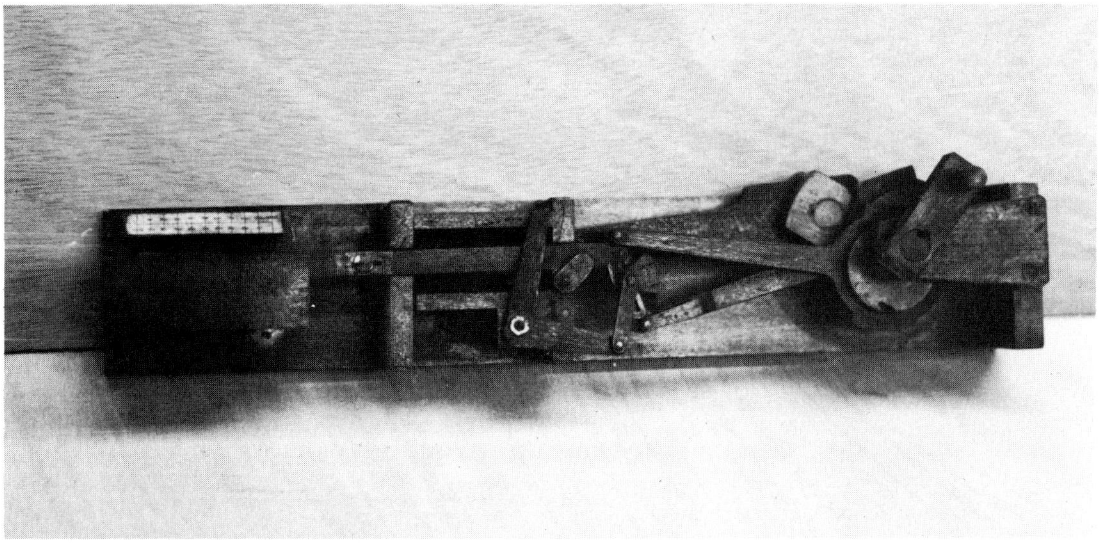

A 'Mattisa' track tamping machine working at Weaste during the construction of the Lancashire Tar Distillers' Oil Compound in 1965.

J. Steggles

Platelayers at work near Weaste Junction in June 1978. Mechanisation arrived late in the day on the Canal and the machine shown in the picture is a Robel Motorized Ganger.

D. Thorpe

anything to avoid having to ask for help when struggling with a train on a gradient. It was a source of some wonderment, and a great deal of amusement, to new firemen when they were put with one of these drivers. To see a driver almost praying by the boiler front for that little bit extra to help them over the top, especially if there was another engine nearby, was not unusual. Today, of course, this pride in the engine is almost dead. It is true that the diesel is more efficient, but there is little or no special skill required to get the best out of the engine, for that has been pre-determined by the makers.

Permanent Way Staff

Permanent way staff were taken on, in the first instance, as labourers and as there were no formal training facilities, learned the job the hard way by simply working alongside the more experienced platelayers. At the present time the permanent way staff number about twenty five and are in two gangs, one being at Ellesmere Port which comprised six men, the remainder being based at Salford. In past years the staff level was in the region of fifty five not including those at Runcorn, and there were three gangs. Each of these gangs was under the control of the Resident Engineer of the section which, until the early 1950s involved three engineers, one over the Manchester Section which ended at Barton Aqueduct, the second over the Irlam Section which extended to Wigg Works, and the third one over the Ellesmere Port/Stanlow Section.

Whilst on the subject of railway lines, a curious custom on the Manchester Ship Canal is the mispronunciation of the word 'route' when referring to the through line. The railwaymen always pronounced it 'rowt' as in shout. How this came about is not really clear but the pronunciation is still the same to this day.

Today there are but a few employed on the railway but there are still plenty of characters and good humour. There is also a certain pride in being part of what was one of the most interesting private railway systems in Britain.

With the introduction of diesel locomotives, the crew was reduced to driver and shunter. In this photograph, showing Hudswell Clarke 204hp diesel-mechanical locomotive No. D14, is shunter Chris McCann and the last of the Agnew family serving the Canal, Raymond.

Courtesy R. Agnew

Manchester Ship Canal locomotive No. D6 face to face with *Flying Scotsman*! This unlikely event occurred when the MSC locomotive was hired out to the organizers of the 'Rocket 150 Exposition', held at the original L&MR Liverpool Rd. Station, Manchester in October 1980. No. D6 was used to move the exhibits in and out of the shed, and was sponsored by the Transport Salaried Staffs Association.

D. Thorpe

Chapter Seven The Locomotives

It is not known for certain just how many locomotives worked on the Canal during the seven year period. The engines which are known to have worked on the site are listed in the Appendix, the list reading like a 'Who's Who' of locomotive builders.

At the time the Company took over the construction, following the death of T. A. Walker, they became the owners of 173 engines. On completion of the Canal, only 23 or 24 were retained by the Manchester Ship Canal for their newly-formed Railway Department. The majority of these were saddle tanks and ranged from a diminutive 0-4-0 with 10in. x 15in. cylinders to an 0-6-0 having 15in. x 20in. cylinders. The name of the 10in. locomotive was *Cheadle* and that of the 15in. locomotive was *Glamorgan*. Unfortunately, there is no record of early stock movements but eye-witness accounts indicate that *Cheadle* was used at Runcorn in the early years and this would probably be correct as she had worked in the section during the construction period.

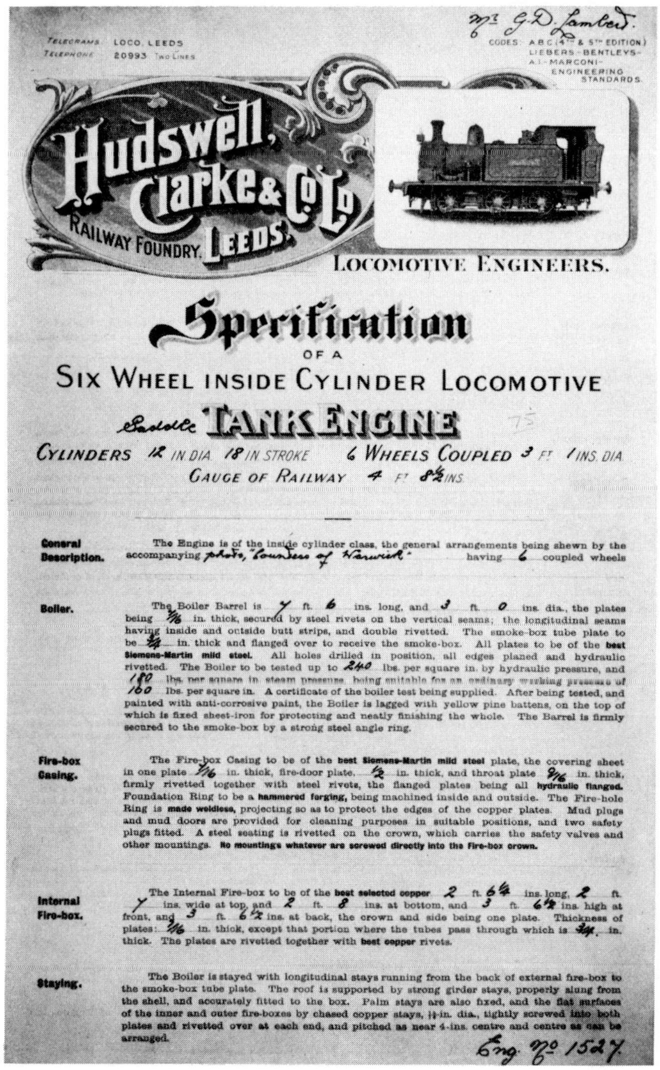
Mr G. D. Lambert.

Telegrams LOCO. LEEDS
Telephone 20993 Two Lines

Codes: A.B.C. (4th & 5th Edition) Liebers - Bentleys - A1 - Marconi - Engineering Standards.

Hudswell, Clarke & Co. Ld. Railway Foundry, Leeds.

Locomotive Engineers.

Specification of a Six Wheel inside Cylinder Locomotive Saddle Tank Engine

Cylinders 12 in. dia. 18 in. stroke 6 Wheels Coupled 3 ft. 1 ins. dia.
Gauge of Railway 4 ft. 8½ ins.

General Description. The Engine is of the inside cylinder class, the general arrangements being shewn by the accompanying photo, "Countess of Warwick" having 6 coupled wheels

Boiler. The Boiler Barrel is 7 ft. 6 ins. long, and 3 ft. 0 ins. dia., the plates being 7/16 in. thick, secured by steel rivets on the vertical seams; the longitudinal seams having inside and outside butt strips, and double rivetted. The smoke-box tube plate to be 3/4 in. thick and flanged over to receive the smoke-box. All plates to be of the **best Siemens-Martin mild steel.** All holes drilled in position, all edges planed and hydraulic rivetted. The Boiler to be tested up to 240 lbs. per square in. by hydraulic pressure, and 180 lbs. per square in. steam pressure, being suitable for an ordinary working pressure of 160 lbs. per square in. A certificate of the boiler test being supplied. After being tested, and painted with anti-corrosive paint, the Boiler is lagged with yellow pine battens, on the top of which is fixed sheet-iron for protecting and neatly finishing the whole. The Barrel is firmly secured to the smoke-box by a strong steel angle ring.

Fire-box Casing. The Fire-box Casing to be of the **best Siemens-Martin mild steel** plate, the covering sheet in one plate 7/16 in. thick, fire-door plate, 1/2 in. thick, and throat plate 9/16 in. thick, firmly rivetted together with steel rivets, the flanged plates being all **hydraulic flanged.** Foundation Ring to be a **hammered forging,** being machined inside and outside. The Fire-hole Ring is **made weldless,** projecting so as to protect the edges of the copper plates. Mud plugs and mud doors are provided for cleaning purposes in suitable positions, and two safety plugs fitted. A steel seating is rivetted on the crown, which carries the safety valves and other mountings. **No mountings whatever are screwed directly into the Fire-box crown.**

Internal Fire-box. The Internal Fire-box to be of the **best selected copper** 2 ft. 6¼ ins. long, 2 ft. 7 ins. wide at top, and 2 ft. 8 ins. at bottom, and 3 ft. 6½ ins. high at front, and 3 ft. 6½ ins. at back, the crown and side being one plate. Thickness of plates: 7/16 in. thick, except that portion where the tubes pass through which is 3/4 in. thick. The plates are rivetted together with **best copper** rivets.

Staying. The Boiler is stayed with longitudinal stays running from the back of external fire-box to the smoke-box tube plate. The roof is supported by strong girder stays, properly slung from the shell, and accurately fitted to the box. Palm stays are also fixed, and the flat surfaces of the inner and outer fire-boxes by chased copper stays, 15/16-in. dia., tightly screwed into both plates and rivetted over at each end, and pitched as near 4-ins. centre and centre as can be arranged.

Eng. No. 1527.

Locomotive specification sheets for No. 75, issued by the makers, Hudswell Clarke & Co. The locomotive shown at the top is *Philadelphia* which is a short tank engine, whilst locomotive No. 75 was a small 0-6-0 saddle tank engine.

D. Thorpe

By 1915 the numbers of ex-contract engines had dwindled to 10 and these formed the basis of the numbered fleet. Locomotives which were numbered 11 to 90, which was the final steam engine bought, totalled 81, the number 34 being taken by two engines at different periods. Oddly enough, both engines were bought second-hand as were the three 'Austerity' engines, but all other locomotives bought by the Company were new.

Almost all the engines taken from the contract by the Company were later sold, whether to other users or whether to scrap dealers it is not possible to say. Records do not show, for instance, what happened to No. 1 *Elham*, (it is supposed to have been scrapped in 1929 and 1933!) but they do show that No. 2 *Romilly* lasted longer on the Manchester Ship Canal than any other ex-contractor's locomotive, being sold by the Company in 1932.

Locomotive No. 5 *Douglas*, was taken by the Mechanical Engineer, Mr Fox, and used as an inspection vehicle. Part of the back of the cab was fitted with a recess and a seat was installed so that Mr Fox could sit in the cab out of the way of the driver and fireman. The locomotive had to be kept even more spick and span than the rest of the fleet, and they were also very well looked after.

One locomotive which was never recorded as being received by the Company during the construction period, and also was not recalled by eye-witnesses, was the 0-4-0 Hudswell Clarke engine, *Lord Mayor*. The engine was, however, delivered to a sub-contractor, Edmund Nuttal, at Salford Docks in 1893, but as Edmund Nuttal was involved in the construction of Manchester Dry Docks, the locomotive could well have been employed on that site.

When the Company commenced buying new locomotives specifically for traffic duties, it continued the practice of naming them but, in keeping with its business as a sea port, it had them named after international sea ports or countries having trade ties with Manchester.

Locomotives purchased specifically for Traffic Duties

As the Manchester Ship Canal grew in importance, the docks became busier, and existing locomotives were being used to full capacity. Consequently, in 1896, two extra locomotives were ordered new from Peckett & Sons Ltd. They were both 0-4-0 saddle tank engines of the W4 class with outside cylinders and, like all Manchester Ship Canal locomotives, were fitted with Stephenson's link motion. Braking was by hand screw only and throughout their lifetime they were never fitted with a steam brake. An annoying feature was the lack of adequate bunkering. Coal had to be piled on to the floor at the back of the cab, which was not only a nuisance but was also dangerous. Even so, there were drivers who liked them very much, and preferred working them to other classes of locomotive.

Hunslet Engines

The following year saw the arrival of two further locomotives but this time from the Hunslet Engine Company. Both engines, which were named after Canadian ports, arrived in

Another of the unnumbered, ex-contractor's engines working for the Railway Department. *Bollin* was built by the Hunslet Engine Co. in 1889 (Works No. 462) and was delivered new to the Canal construction site.

M S C Co.

Locomotive No. 1 *Elham*, photographed at No. 1 Grid, Salford, circa 1931. Built in 1884 by the Hunslet Engine Co. (Works No. 353), the engine was intended for the Canterbury to Shorncliffe Contract (*Elham* being a place name in that area) but was first delivered to Preston where T. A. Walker was contracted to build the new docks system. Eventually, it arrived on the Manchester Ship Canal Contract where it stayed even after the construction was completed, forming part of the Company's locomotive fleet. It remained until at least 1929, possibly 1933, the exact disposal date being unrecorded. Note the dumb buffers still fastened to the buffer beam.

Courtesy A. Appleton

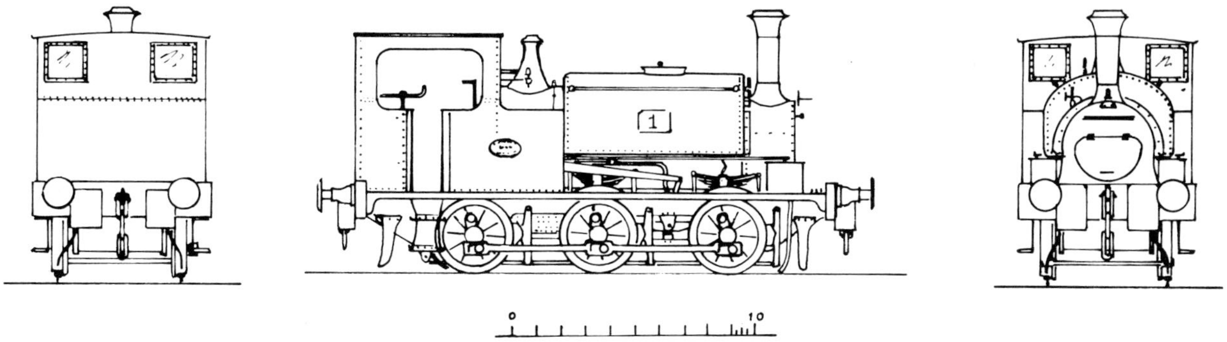

Maker	Hunslet Engine Co., Leeds
Type	0-6-0 Contractor's engine
Other engines in same class	
Length over buffer beams	19ft. 7in.
Length overall	
Maximum width	7ft. 3½in.
Height over chimney	11ft.
Cylinders	12in. x 20in. (inside)
Wheel diameters	3ft. 3in. over tyre
Wheelbase	10ft. 6in. (L to D 5ft. 4in.) (D to T 5ft. 2in.)
Weight (empty)	
Weight (in working order)	20tons 0cwt.
Capacity (coal)	
Capacity (water)	500 galls. (imperial)
Boiler pressure	
Heating surface (total)	
Heating surface (tubes)	
Heating surface (firebox)	
Grate area	
Tractive effort	

Locomotive No. 2, *Romilly*, again a Hunslet engine and also an ex-contract engine, stands outside Mode Wheel locomotive shed, circa 1931, together with Hudswell Clarke locomotive No. 79 and others. It was delivered new in 1885 (Works No. 357) to Cardiff where it was put to work on the Barry Docks Contract for T. A. Walker. On being transferred to the Manchester Ship Canal contract, it was put to work at Irlam. After being transferred to the Company's fleet in 1894 the engine worked on for a further 38 years before being sold in 1932, but whether for further use or whether for scrap is not known. The photograph shows clearly the high standard of cleanliness maintained at that time with handrails and buffers being polished, as was the brasswork and paintwork.

Courtesy A. Appleton

Ex-contract engine No. 6 *Longsight* at Partington. Built in 1888 by the Hunslet Engine Co., (Works No. 455), the engine came new to T. A. Walker at Manchester and remained on the Canal until sold in 1931.

Courtesy J. Steggles

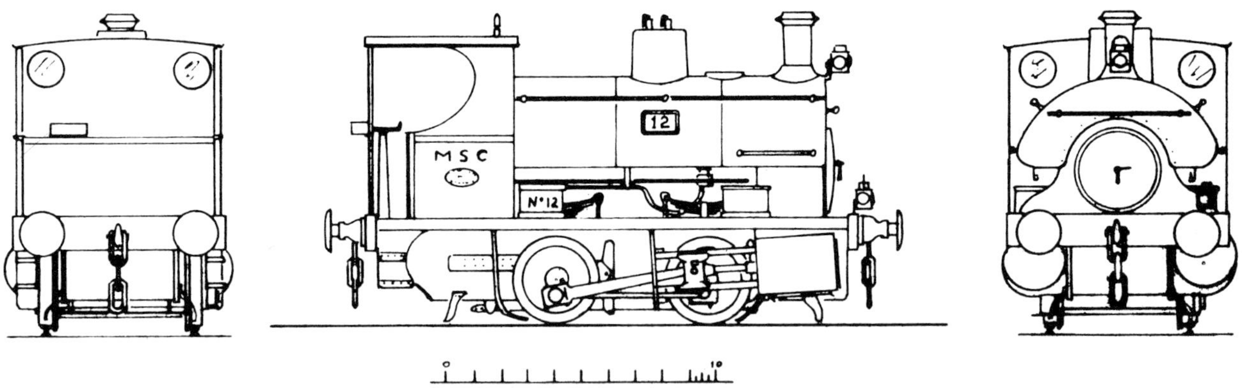

Maker	Peckett & Sons, Bristol
Type	Class 4 (0-4-0 saddle tank)
Other engines in same class	Loco. 11 and (with modifications) Loco. 25
Length over buffer beams	
Length overall	18ft.
Maximum width	8ft.
Height over chimney	11ft. 7in.
Cylinders	14in. x 20in. (outside)
Wheel diameters	3ft. 3½in.
Wheelbase	5ft. 6in.
Weight (empty)	18 tons 12 cwt.
Weight (in working order)	23 tons
Capacity (coal)	1 ton 9 cwt.
Capacity (water)	660 galls.
Boiler pressure	
Heating surface (total)	
Heating surface (tubes)	
Heating surface (firebox)	
Grate area	
Tractive effort	

December, *Montreal* being delivered first, followed by *St. John* one week later.

Locomotive	*Maker*	*Arrived*
Montreal	Hunslet Engine Co.	2.12.1898
St. John	Hunslet Engine Co.	9.12.1898

To quote the Half-Yearly Company Report, 'they were of a special more powerful type'. Although they were not large engines, they were certainly heavier and more powerful than most of their predecssors. An interesting feature of the design was the use of the Cartazzi leading axleboxes. This, added to the vertical knuckle pins incorporated into the forward coupling rods, gave extra flexibility to the locomotive. Another feature, which increased the engines ability to negotiate short chain curves, was the absence of flanges on the driving wheels.

The Cartazzi system was simply a self-centring device, on the leading axleboxes, which used the action of a series of inclined faces machined on the top of the axlebox and on the underside of a very substantial axlebox lid. The lid, which was $1^3/_8$in. thick, had three sets of shallow-angled vees, the extreme end areas sloping inwards to the centre line, the middle area forming a peak. At the top of the axlebox were three vee areas which formed matching surfaces. In this way the axlebox, when forced sideways, would have a three point seating with the lid, on faces which were inclined the same way. The lid, which was subject to the downward pressure of the bearing spring, was a close fit in the horns unlike the

The boiler front in the cab of a 'Jazzer'.

B. W. Potts

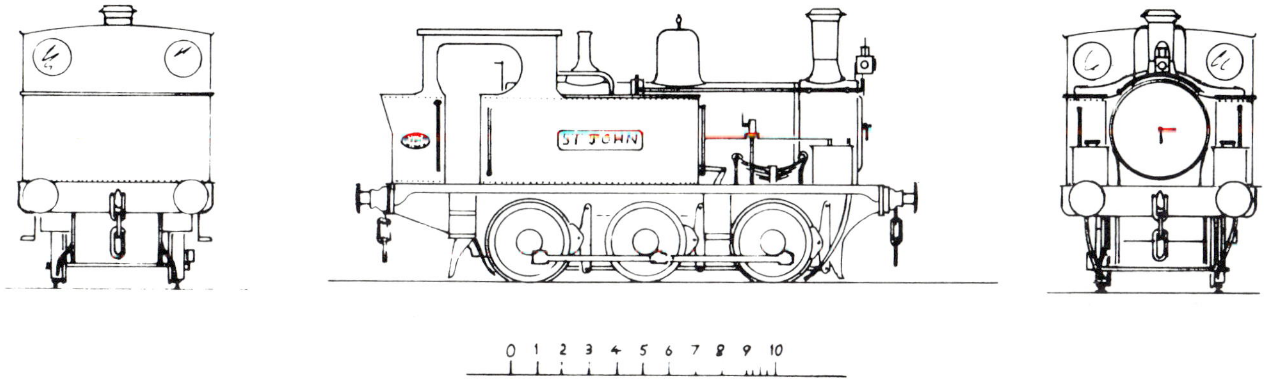

Maker	Hunslet Engine Co., Leeds
Type	0-6-0 side tank (Code 'Chest') (MSC 'Jazzer')
Other engines in same class	Total 10: Loco. 13 to Loco. 22
Length over buffer beams	21ft. 0in.
Length overall	24ft. 0in.
Maximum width	8ft. 5in.
Height over chimney	11ft. 6in.
Cylinders	15½in. x 20in. (inside)
Wheel diameters	3ft. 4in.
Wheelbase	10ft. 3in. (L to D 5ft. 4in.) (D to T 4ft. 11in.)
Weight (empty)	24 tons 3 cwt.
Weight (in working order)	29 tons 17 cwt.
Capacity (coal)	1 ton 5 cwt.
Capacity (water)	600 galls.
Boiler pressure	160 p.s.i.
Heating surface (total)	706ft. (squared)
Heating surface (tubes)	647ft. (squared)
Heating surface (firebox)	59ft. (squared)
Grate area	10.4ft. (squared)
Tractive effort	14,410lb. at 75% boiler pressure

axlebox which had a considerable amount of side play. It will be seen that when the leading wheels were forced across on a sharp curve, the tendency was for the lid to be forced upwards against the pressure of the spring. When the locomotive came to run on a straight section the bearing spring would assert itself, and through the action of the inclined faces just described, would bring the axlebox back to a central position. The class proved to be successful, having the ability to produce an extra bit of effort at a speed that would have stalled the later Hudswell Clarke locomotives of similar size.

Their main faults were the awkwardness of firing due to the lack of space between the boiler front and the bunker, and their tendency to jig all over the place when they were taking a heavy train. It was this latter trait which led to them being labelled 'Jazzers', and to this day they are referred to by that name. Braking was by steam and hand screw, whilst the sanders were gravity fed serving the front of the leading wheels and the rear of the trailing wheels only.

The first engine purchased new by the Manchester Ship Canal, specifically for the use of the infant Railway Department, Peckett & Son locomotive No. 654 of 1897, *Alexandria* (No. 11) was seen standing by the coal stage at Mode Wheel yard. It was a particularly useful locomotive especially at Pomona and on the even tighter system around Ordsall and the abandoned No. 5 Dock. However, in 1954 *Alex*, as she was called, was sold to the Esso Petroleum Co. on Trafford Wharf, lasting a further two or three years before being scrapped, due to expensive boiler trouble.

Alex Appleton

The Railway Department was obviously satisfied with the 'Jazzers', for in 1899 a further two orders, each for two locomotives, was given to the Hunslet Engine Company. 1900 and 1902 saw another four locomotives of the same class, the final one being *Rotterdam*.

Locomotive	*Maker*	*Arrived*
New York	Hunslet Engine Co.	3.8.1899
Galveston	Hunslet Engine Co.	8.9.1899
New Orleans	Hunslet Engine Co.	4.12.1899
Savannah	Hunslet Engine Co.	13.12.1899
London	Hunslet Engine Co.	7.6.1900
Glasgow	Hunslet Engine Co.	31.10.1900

Locomotive	*Maker*	*Arrived*
Antwerp	Hunslet Engine Co.	30.4.1902
Rotterdam	Hunslet Engine Co.	30.4.1902

Hudswell Clarke Locomotives

Why the Manchester Ship Canal stopped ordering from the Hunslet Engine Company is not known but, whilst still in 1902, the Mechanical Engineer's Department took delivery of three more engines, two from Hudswell Clark & Company of Leeds, and one from Peckett & Sons, Bristol.

In their specification the Hudswell Clarke locomotives were very similar to the 'Jazzers', being 0-6-0 side tank engines, and like their Hunslet counterparts, were fitted with Cartazzi leading axleboxes. The coupling rods pivoted at a point just in front of the driving wheel crank pin in the same manner as those on the 'Jazzers'. Even the bearing springs, including the compensating beams, were the same, but after that all similarities ceased. The appearance of the engines was a lot different, the 'Jazzers' having tanks with rounded corners at the front end and a rather severe line to the cab and bunker. All the rivets on the upper part of the engine were round-headed. The Hudswell Clarke locomotives were a lot more pleasing to the eye having curved lines instead of harsh angles. The tanks, bunker, and cab were fastened together using countersunk rivets, the whole being filled and smoothed off before painting. Even the dome and sandboxes were of a more rounded appearance.

The gross weight of the Hudswell Clarke engines was slightly more than those of the Hunslet Engine Company, being 32 tons 14 cwt. With a bunker capacity of 1 ton 10 cwt, the engine carried 20 per cent more fuel than the 'Jazzers' but

An overcast afternoon outside Weaste Cemetery, Salford, with No. 18 *Savannah* and crew, posed for the picture. Built in 1899 by the Hunslet Engine Co., Leeds (Works No. 712), the engine shows the style of lining-in on the tanks, cab and boiler quite clearly. The single safety valve was subsequently replaced by two Ross pop valves.

Courtesy A. Appleton

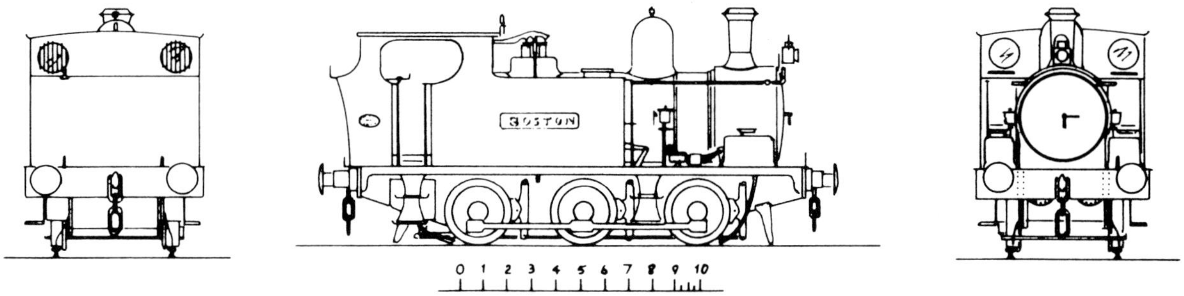

Maker	Hudswell Clarke & Co. Leeds
Type	0-6-0 short tank ('Canal')
Other engines in same class	26
Length over buffer beams	21ft. 0in.
Length overall	24ft. 0in.
Maximum width	8ft. 3in.
Height over chimney	11ft. 6in.
Cylinders	15½in. x 20in. (inside)
Wheel diameters	3ft. 4½in.
Wheelbase	10ft. 3in. (L to D 5ft. 4in.) (D to T 4ft. 11in.)
Weight (empty)	28 tons 16 cwt.
Weight (in working order)	32 tons 14 cwt.
Capacity (coal)	1 ton 10 cwt.
Capacity (water)	580 galls.
Boiler pressure	140 p.s.i.
Heating surface (total)	706ft. (squared)
Heating surface (tubes)	647 ft. (squared)
Heating surface (firebox)	59 ft. (squared)
Grate area	10.4 ft. (squared)
Tractive effort	

A highly-polished *Boston*, posed between the fire droppings and the coal stage at Mode Wheel, some time prior to 1915 and possibly in 1902 when the engine was built. The locomotive was a Hudswell Clarke engine, (Works No. 641). The Ramsbottom type safety valves were later changed to Ross pop valves.

Courtesy B. Valentine

carried 3 per cent less water in tanks, which held 580 gallons. Braking was by steam and hand-operated screw. The steam brake operating valve was fitted with a ratchet and when the valve became worn the handle used to rattle violently. Another part of the brake system which caused bother was the exhaust. This went through the roof and when the brake was realeased, the sudden hiss of escaping steam was rather startling. Whilst working around the docks, the locomotives would often be near the cartroads where there would usually be horse-drawn vehicles standing in line awaiting their loads. The steam brake exhaust often caused the horses to shie violently and this led to angry complaints by the carters. Eventually the brake exhaust was led through a tube downwards to a horizontally-placed perforated tube which lay across the top of the ashpan. For obvious reasons this was called a 'degger' pipe and it directed the steam into the ashes, dampening the noise as well as the ashes. At a later time a new type of steam brake valve was fitted, the design of which was attributed to Mr 'Billy' Lewis, the then Locomotive Superintendent.

It is recorded that when they were steamed for trial at Salford, the representative from Hudswell Clarke had 50 loaded wagons hooked on to the locomotive, then added another one 'to make sure', whereupon the engine pulled them away quite easily.

The maker's plate on locomotive No. 61 *Reval.*

D. Thorpe

Locomotive	*Maker*	*Arrived*
Boston	Hudswell Clarke & Co.	18.12.1902
Philadelphia	Hudswell Clarke & Co.	31.12.1902
Valencia	Peckett & Son	13.3.1902

The Peckett purchased in 1902 was again a Class W4 engine, but with slight modifications, including an altered cab and the fitting of a steam brake. However, the length of the cab was little changed from the earlier Peckett locomotives, the fireman still having a job to fire the boiler, and the storage of coal was still rather poor. *Valencia* was the name given to the new locomotive.

Another 0-4-0 saddle tank arrived at Salford Docks the following year, the maker this time being Hudswell Clarke & Co. When the engine came to the Canal it carried the name *Leeds* and as every Yorkshireman knows, Leeds is not a sea port. The reason for this apparent departure, by the Manchester Ship Canal, from their usual practice of choosing names, is easily explained, for this engine was on hire from the makers for a period of 15 months from 27th July 1903, the date on which she was delivered. When the Company eventually bought the engine they decided to change the name. On 18th October 1904, for the sum of £650 she became a Manchester Ship Canal locomotive and was renamed *Bristol* and she was usually to be seen around Pomona, Ordsall and Salford Yard.

A further, and more surprising, change in Company policy occurred with the purchase from Hudswell Clarke & Co., of a twenty five year old, reconditioned 0-6-0 saddle tank engine, which, when delivered to the Manchester Ship Canal in 1904, bore the name *Copenhagen*. Built in 1880 at the Railway Foundry, Leeds, and carrying the works number 218, it was the last engine to be built by the company under its title, Hudswell, Clarke & Rodgers. The locomotive left the works as No. 6 for the colliery owners J. & J. Charlesworth, and was delivered to Rothwell Haigh Colliery. It was the largest saddle tank engine to be seen on Manchester Ship Canal lines for many years being, at 24 tons 10 cwt, just 4 tons 5 cwt heavier than the last Peckett locomotive delivered. Apart from its weight, the engine looked bigger, not because of its length, which was 22ft. 9½in., but more because of its height. At 11ft. 11½in. to the top of the chimney it was taller than any of the other classes, and this high appearance was further exaggerated by the bulky saddle tank, the high framing and the high slung cylinders. However, the engine worked well and was well liked, being handled by crews at Partington, Acton Grange and Runcorn. Apart from the two saddle tank engines just discussed, there were six Hudswell Clarke short tank locomotives delivered in 1903 and two more in 1904. Also in 1904 the Locomotive Department purchased what was, for the Manchester Ship Canal, a really large engine, an 0-6-0 Kitson.

Locomotive	*Maker*	*Arrived*
Bristol	Hudswell Clarke & Co.	29.7.1903 (as *Leeds*)
Dublin	Hudswell Clarke & Co.	16.1.1903
Belfast	Hudswell Clarke & Co.	23.1.1903
Bombay	Hudswell Clarke & Co.	20.5.1903
Kurrachee	Hudswell Clarke & Co.	27.5.1903
Hamburg	Hudswell Clarke & Co.	25.9.1903
Gothenburg	Hudswell Clarke & Co.	28.9.1903
Christiania	Hudswell Clarke & Co.	18.8.1904
Copenhagen	Hudswell Clarke & Rodgers	11.2.1904
Stockholm	Hudswell Clarke & Co.	17.9.1904
Port Limon	Kitson & Co.	3.12.1904

The First Manchester Ship Canal Kitson Locomotive

The last locomotive to be delivered in 1904 arrived on 3rd December from the makers, Kitson & Company Leeds. When delivered, the driving wheels were flangeless in the fashion of the 'Jazzers' and the Hudswell Clarke short and long tank locomotives. Unfortunately, with such a long wheelbase, the driving wheels tended to drop off the rails when the engine was working a section which included short radius curves. Eventually it was retyred, all wheels being flanged. On delivery, the Kitson, which was named *Port Limon*, was put to work at Partington (north).

From the point of view of the fireman and driver, the Kitson was a mixed blessing. With the larger firebox to be fed and the bigger usage of steam, it is obvious that the fireman had to work harder, but he did have more room to move around in, the cab being so spacious in comparison with that of previous engines. However, even the extra space had its drawbacks, for in the winter the cab could be really cold, and that was in spite of the fact that there were half doors fitted and sliding plates

This photograph, taken possibly circa 1906, shows a very smart *Valencia* (later MSC No. 25), at Salford. Built by Peckett & Son in 1902 (Works No. 907) the engine spent most of its years working around Ordsall and Pomona with a few years at Runcorn working for Bridgewater Department's Railway.

M S C Co.

Hudswell Clarke & Rodgers engine, MSC No. 34 *Copenhagen*, on the gradient at Acton Grange.

Courtesy A. Appleton

Locomotive No. 36 *Port Limon* at Partington North on 18th June 1952, where she worked for most of her days. Bought in 1904, this was the first Kitson engine purchased new by the Company for traffic purposes.

A. Appleton

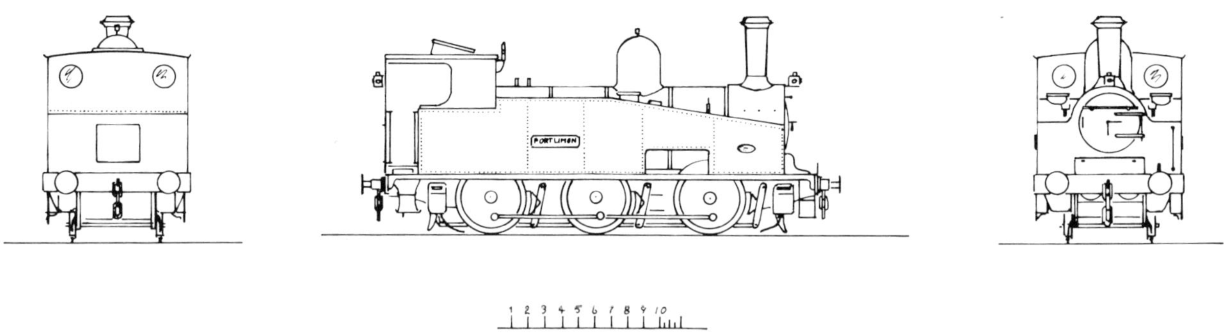

Maker	Kitson & Co., Leeds
Type	0-6-0 side tank. *Port Limon* (No. 36)
Other engines in same class	Total 4: Locos. 36, 72, 73 & 74
Length over buffer beams	25 ft. 10 in.
Length overall	29 ft. 10 in.
Maximum width	8 ft. 9 in.
Height over chimney	13 ft.
Cylinders	18 in. x 26 in. (inside)
Wheel diameters	4 ft. 6 in.
Wheelbase	13 ft. (L to D 6 ft. 9 in.) (D to T 6 ft. 3 in.)
Weight (empty)	35 tons 14 cwt.
Weight (in working order)	44 tons 7 cwt.
Capacity (coal)	1 ton 4 cwt.
Capacity (water)	1,350 galls.
Boiler pressure	180 p.s.i.
Heating surface (total)	927.84 ft. (squared)
Heating surface (tubes)	864.29 ft. (squared)
Heating surface (firebox)	82.84 ft. (squared)
Grate area	18.7 ft. (squared)
Tractive effort	21,060 lb.

on each side of the cab. During the hot weather, the sides of the cab could be opened up to a certain extent and the hatch on the cab roof also opened, making working conditions a little more bearable.

Driving the locomotive was really a job for a man of 7ft. in height and as these are hard to come by, a set of raised wooden platforms was fitted, one on either side of the boiler front. The regulator was of the single lever type, like those of previous engines, but the fire hole door was completely different in that it hinged downwards to open. Braking was by steam and hand screw. Sanders were gravity fed and operated on front and rear wheels only. Leaf springs were fitted on each axlebox, driving and trailing springs being underslung, whilst the leading springs were set on spring pillars above the axleboxes. On each pair of horns was fitted an adjustable wedge-shaped horn block. This very useful device was used to take up any slack which appeared between the axlebox liners and the horn cheeks. A single nut on a stud at the back of the horn block would be released, after which the adjusting screw, fitted to the underside of the block, would be adjusted, thrusting the wedge upwards. Finally the back nut would be retightened. It was a simple, but effective, device, and accounted, in part, for the longer life of the axleboxes and running gear when compared with that of other locomotives.

Although *Port Limon* was a successful addition to the locomotive stud, another seventeen years was to pass before the department purchased a second similar engine from Kitson & Company.

In the years between, the only locomotive builder to receive orders from the Manchester Ship Canal Company was Hudswell Clarke & Company. In that period they supplied a further 14 short tank and twenty long tank locomotives of the 'Sweden' class. During 1905 there were three short tank engines delivered, their names being *Adelaide, Melbourne* and *Sydney*. With the opening in 1905 of No. 9 Dock, and the 1906 Agreement between Trafford Park Estates and the Manchester Ship Canal Company, making the Company agents for all railway traffic between firms in the 'Park', and the lines of the railway companies, the Railway Department needed extra engines to cope with the expected increase in traffic. Subsequently a total of six short tank locomotives arrived during 1906.

Locomotive	*Maker*	*Date Delivered*
Adelaide	Hudswell Clarke & Co.	12.8.1905
Melbourne	Hudswell Clarke & Co.	29.8.1905
Sydney	Hudswell Clarke & Co.	1.9.1905
Buenos Ayres	Hudswell Clarke & Co.	17.2.1906
Monte Video	Hudswell Clarke & Co.	23.2.1906
Rossario	Hudswell Clarke & Co.	7.3.1906
Wellington	Hudswell Clarke & Co.	21.6.1906
Auckland	Hudswell Clarke & Co.	27.7.1906
Jamaica	Hudswell Clarke & Co.	27.10.1906

Another good order followed for a further six short tank locomotives which were delivered in 1907. Their names followed the pattern of the three previous engines, in that they were named after countries and ports of the Empire, or Commonwealth as it is today, all but one that is, for *Stettin* was the odd one out.

Locomotive	*Maker*	*Arrived*
Quebec	Hudswell Clarke & Co.	5.1.1907
Canada	Hudswell Clarke & Co.	22.1.1907
India	Hudswell Clarke & Co.	12.2.1907
Stettin	Hudswell Clarke & Co.	31.3.1907
Australia	Hudswell Clarke & Co.	19.4.1907
New Zealand	Hudswell Clarke & Co.	30.4.1907

For the next 4½ years, no locomotives were bought. One of the few recorded changes in the locomotive stock position

The locomotive *Kurrachee* with members of both Railway and Locomotive Departments. On the footplate is Driver G. Beard, whilst at ground level is Fireman J. McNeil, Shunter W. Robinson and, at the front end, Foreman Shunter H. Emery. The photograph, taken outside Mode Wheel Workshops, on the dead-end, shows the engine when she was about three years old. A Hudswell Clarke product, the engine was built in 1903 (Works No. 663), and later became MSC No. 30.

Courtesy M S C Co.

was the sale of *Severn*, sometime during 1910. That is not to say that she was the only ex-contract engine sold during this period, for several of them are merely recorded as being sold but at what date and to whom there is no indication.

Long Tank Hudswell Clarke Locomotives

With the advent of the Irlam Steelworks of the Partington Steel & Iron Company, the increased heavy traffic created the need for locomotives which were suitable for train running. Of course, the Kitson, *Port Limon* was well able to handle large trains, but she was more useful on the gradient from the coaling basin up to Glazebrook Junction than any other of the Company's locomotives. The Hudswell Clarke short tank engines could also take quite sizeable trains, especially on the comparatively level route between Partington and Latchford, the latter being the nearest point at which the Manchester Ship Canal's lines in that section connected with those of the LNWR.

The main drawback of the Hudswell Clarke short tank engines for work of this nature was their small water capacity. Accordingly, the makers modified the design by extending the tanks to a point level with the front of the smokebox. This increased the water capacity to 840 gallons, the gross weight of the locomotive when full, increasing to 33 tons 19 cwt, an increase of 260 gallons and 1¼ tons above the short tank locomotives. Although the makers referred to them as the 'Sweden' type, the Canal men always called them simply, 'long tankers'.

The following year, the Company placed an order for another three. In order of maker's number and delivery, the names were *Odessa*, *Batoum* and *Smyrna*. Again the livery was light green with black and white lining.

The class was very successful and in 1913 a further six were ordered, four arriving in that year and two arriving in 1914.

Locomotive	*Maker*	*Arrived*	*First Duty*
Norway	Hudswell Clarke & Co.	20.10.1911	
Sweden	*Hudswell Clarke & Co.*	*Dec. 1911*	
Odessa	Hudswell Clarke & Co.	Oct. 1912	
Batoum	Hudswell Clarke & Co.	Dec. 1912	
Smyrna	Hudswell Clarke & Co.	6.2.1913	
Dunkirk	Hudswell Clarke & Co.	2.10.1913	
Bordeaux	Hudswell Clarke & Co.	31.10.1913	Partington 1.11.1913
Ghent	Hudswell Clarke & Co.	6.11.1913	
Riga	Hudswell Clarke & Co.	21.12.1913	
Reval	Hudswell Clarke & Co.	11.2.1914	
St. Petersburg	Hudswell Clarke & Co.	25.2.1914	Partington 28.2.1914

Locomotive No. 69, Hudswell Clarke long tank, (Works No. 1464), built in 1921, with Driver Harry Harden and Fireman Alf Blanchard (later to become Locomotive Superintendent) standing on No. 2 Grid. To the engine builders, the type was referred to as the 'Sweden' class. The livery was in the original style, but where the earlier engines were light green with black edging and white lining, this was one of the post-World War I engines which were painted a slate blue (or grey) colour.
Courtesy Mrs Loddington

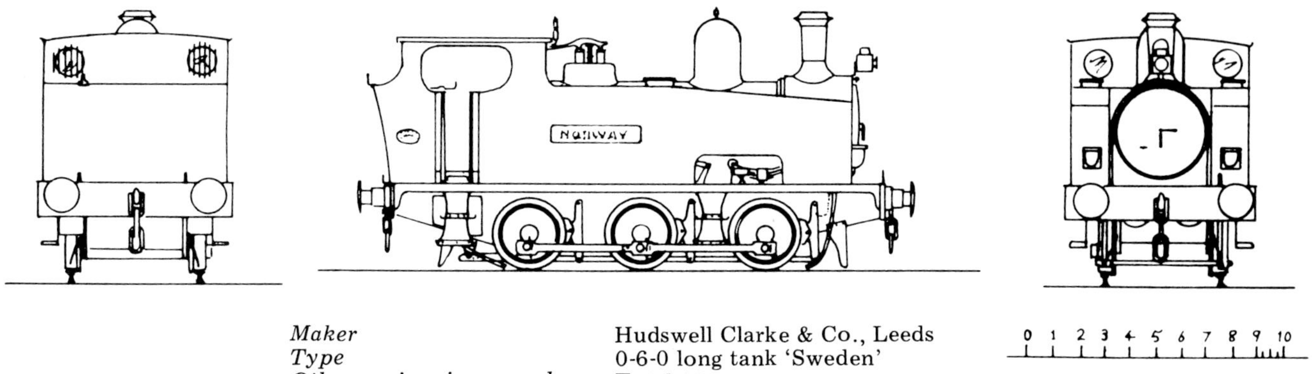

Maker	Hudswell Clarke & Co., Leeds
Type	0-6-0 long tank 'Sweden'
Other engines in same class	Total 22:
Length over buffer beams	21ft.
Length overall	24ft.
Maximum width	8ft. 3in.
Height over chimney	11ft. 6in.
Cylinders	15½in. x 20in. (inside)
Wheel diameters	3ft. 4½in.
Wheelbase	10ft. 3in. (L to D 5ft. 4in.) (D to T 4ft. 11in.)
Weight (empty)	29tons 0cwt.
Weight (in working order)	33tons 19cwt.
Capacity (coal)	1ton 10cwt. 5.5 cu. ft.
Capacity (water)	840 galls.
Boiler pressure	140 p.s.i.
Heating surface (total)	707 sq. ft.
Heating surface (tubes)	647 sq. ft.
Heating surface (firebox)	59 sq. ft.
Grate area	10.4 sq. ft.
Tractive effort	

Manning Wardle No. 1066, MSC No. 7 *Bolton,* with Hudswell Clarke locomotive No. 40, *Buenos Ayres* (Works No. 755) of 1906, posed on No. 1 Grid, Salford. No. 7 was built in 1888 and was delivered new to T. A. Walker at Manchester. It is not recorded where or when the engine was disposed of but it was certainly after 1931 when this photograph was taken.

Courtesy A. Appleton

Numbering of The Locomotive Fleet

During the 1914/15 period all the Company's locomotives were given numbers, the remaining ex-contractor's engines, by that time ten in number, being given the first numbers which started at 1. Brass numberplates were bolted on the tank sides in place of nameplates, one of which was retained and bolted inside the cab on the rear plate. At that time the Company owned 62 locomotives. From the clerk's point of view, the change from names to numbers must have come as a relief, for it must have been a time consuming task making out the locomotive roster. The laborious business of writing names such as *Christiania, Buenos Ayres, Kurrachee,* and others like them must have made the job take ten times as long as it did when the numbering system came into use.

Change of Livery

After the end of World War I, in 1918, the difficulty of obtaining new locomotives lessened and the Company promptly ordered four more long tank engines, but it was March 1919 before the first one was delivered. This engine, locomotive 65, and all subsequent new engines bought before 1954, were turned out in a new livery. Where the earlier colour had been green, albeit different shades, the new colour was grey, or, as the Mechanical Engineer and the makers would have it, slate blue.

A further three were delivered in 1921, one of these, locomotive 69, having been ordered the previous year. Also delivered in 1921 was a Kitson locomotive, No. 72, which was of almost identical appearance to *Port Limon* which, by then, had become No. 36. Although it was not really noticeable, locomotive No. 72 was 4in. shorter in length than the earlier Kitson locomotive but this only affected the overhang at each end, the wheelbase and tank length remaining the same.

No. 72 was sent to Partington three days after arrival at Salford.

Locomotive	*Maker*	*Arrived*	*First Duty*
63	Hudswell Clarke & Co.	26.11.1916	
64	Hudswell Clarke & Co.	1.12.1916	
65	Hudswell Clarke & Co.	1.3.1919	
66	Hudswell Clarke & Co.	18.4.1919	Partington 25.4.1919
67	Hudswell Clarke & Co.	7.5.1919	
68	Hudswell Clarke & Co.	2.6.1919	

Locomotive	*Maker*	*Arrived*	*First Duty*
69	Hudswell Clarke & Co.	4.3.1921	
70	Hudswell Clarke & Co.	30.12.1921	
71	Hudswell Clarke & Co.	30.12.1921	
72	Kitson & Co.	21.10.1921	Partington 24.10.1921
73	Kitson & Co.	12.9.1923	Partington 17.9.1923
74	Kitson & Co.	18.9.1923	Partington 22.9.1923

The success of the Kitson locomotive at Partington ensured further orders for the makers, the next two being delivered in 1923. No. 73 arrived on 12th September and after the usual checks had been carried out at Mode Wheel shops, it was sent to Partington on 17th September. No. 74 was delivered on the 18th and was sent to the coaling basin on 22nd September.

At the Ellesmere Port Docks section it had been realized that the Hudswell Clarke long tank engines were rather large for certain sections, and consequently a smaller type of engine

MSC locomotive No. 8 *Heywood*, Manning Wardle No. 1082, built in 1888, was delivered new to the Canal contract at Salford where it continued to work. It was of a larger class than most contractor's engines and was a natural choice for a position in the Company's early locomotive fleet. The photograph was taken at Weaste before the engine was sold by the Company in 1927.

Courtesy A. Appleton

Hudswell Clarke engine No. 75, built in 1924, (Works No. 1527), seen at No. 1 Grid at Salford, circa 1930. Although this engine and its sister engine, No. 78, performed their task ably, the number of duties needing engines of their size were few; consequently they were, for most of their years, rather under-employed. However, in 1938 both engines were transferred to the Bridgewater Department's Railway at Runcorn Docks where they proved very useful. When No. 78 was transferred back to Salford in 1952 the fitters at Mode Wheel shops, who were more familiar with engines of a more tidy appearance, likened the design to that of a chip shop range and referred to No. 78 as the 'Fish & Chip' engine.

Courtesy Alex Appleton

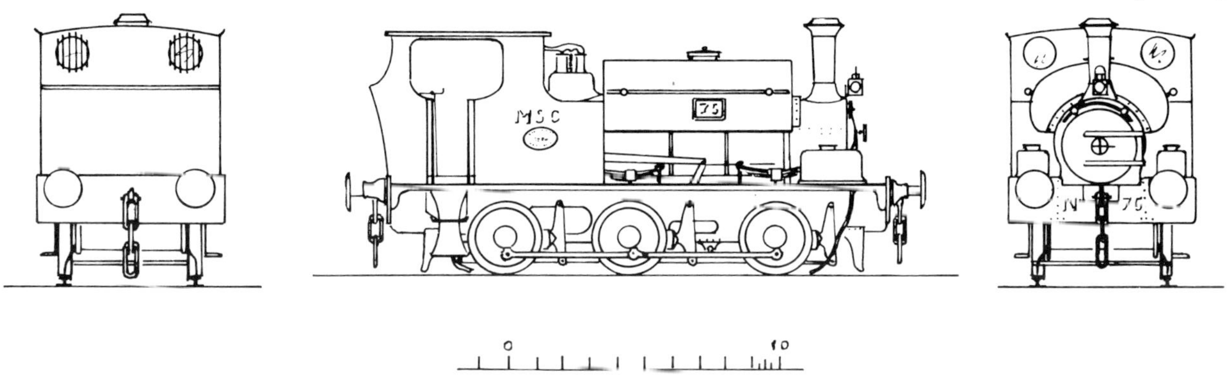

Maker	Hudswell Clarke & Co., Leeds
Type	0-6-0 saddle tank
Other engines in same class	Total 2: Locomotives 75 & 78
Length over buffer beams	19ft. 7in.
Length overall	21ft. 1in.
Maximum width	7ft. 5½in.
Height over chimney	10ft. 3in.
Cylinders	12in. x 18in. (inside)
Wheel diameters	3ft. 1in.
Wheelbase	10ft. 6in.
Weight (empty)	18tons
Weight (in working order)	22tons
Capacity (coal)	13½cwt.
Capacity (water)	500 galls.
Boiler pressure	160 p.s.i.
Heating surface (total)	382.7 sq. ft.
Heating surface (tubes)	341.7 sq. ft.
Heating surface (firebox)	41 sq. ft.
Grate area	6.3 sq. ft.
Tractive effort	

was ordered from the same makers. The new locomotive, No. 75, was a 12in. 0-6-0 saddle tank with a gross weight of 22 tons. Although the overall length at 21ft. 2in. was almost 4ft. shorter than the two previous classes of Hudswell Clarke engines, the wheelbase was longer by 3in. It is surprising that the Company chose this type of engine, for apart from the excessive weight of the earlier locomotives, another reason for having a different type was the difficulty experienced by the previous engines in negotiating certain tight curves in the old dock area. The fact that all wheels were flanged on No. 75 did not improve matters.

Apparently the locomotive was not a success at Ellesmere Port and after a time was transferred to Salford where it was found to be more useful. It was a good steamer and whilst at Salford must have redeemed itself, for another of the same class was ordered three years later.

To return to the period between the purchases of the two engines, No. 76, another long tank engine was delivered on 29th January 1927 and promptly earned the name of *Secotine*, the brand name of a type of glue, because she tended to 'stick' when stopped in a certain position. After adjustment to the valve gear the trouble was eliminated.

A month after the arrival of No. 76, No. 77 was delivered. The new engine, a Class W6 Peckett saddle tank, was larger than the three previous Peckett locomotives. Outwardly it was of much chunkier apperance with a larger tank, a shorter projecting chimney, and a wrap-round back plate on the cab. Coal was carried in two boxes, one on each side of the firebox, at the front part of the cab.

With a larger boiler holding 124 tubes of 1¾in. x 9ft. 8½in. diameter, which were 9in. longer than those on earlier engines, the steaming capability was very good. This, and the 22in. stroke of the pistons, gave the engine a useful increase in drawbar pull. No. 77 spent most of its life in Pomona and Ordsall. The other Hudswell Clarke saddle tank, No. 78, was purchased by the Mechanical Engineer for the Salford Resident Engineer's Department, and it arrived in March 1927. One of its tasks was the shunting of 'muck trains'; these being used to take the rubbish, swept up from around the quay and sidings, along to the rubbish tips at Statham. In that year, the Company disposed of two more ex-contract engines, No. 8 *Heywood* and No. 10 *Irk*. One of the Peckett locomotives, No. 12 *Jaffa*, was also sold. The Manchester Ship Canal locomotive stud then totalled 75.

Locomotive	*Maker*	*Arrived*	*First Duty*
75	Hudswell Clarke & Co.	10.10.1924	Ellesmere Port
76	Hudswell Clarke & Co.	29.1.1927	
77	Peckett & Son Ltd.	24.2.1927	
78	Hudswell Clarke & Co.	26.3.1927	
79	Hudswell Clarke & Co.	15.6.1927	
80	Kitson & Co.	16.4.1927	

Locomotive	*Maker*	*Arrived*	*First Duty*
81	Kitson & Co.	12.1.1929	
82	Kitson & Co.	17.1.1929	

A very clear illustration of the lining-in of locomotive No. 77, the Class W6 Peckett. Built in 1927, (Works No. 1724), the engine spent nearly all its life in Ordsall and Salford yard.

Courtesy A. Appleton

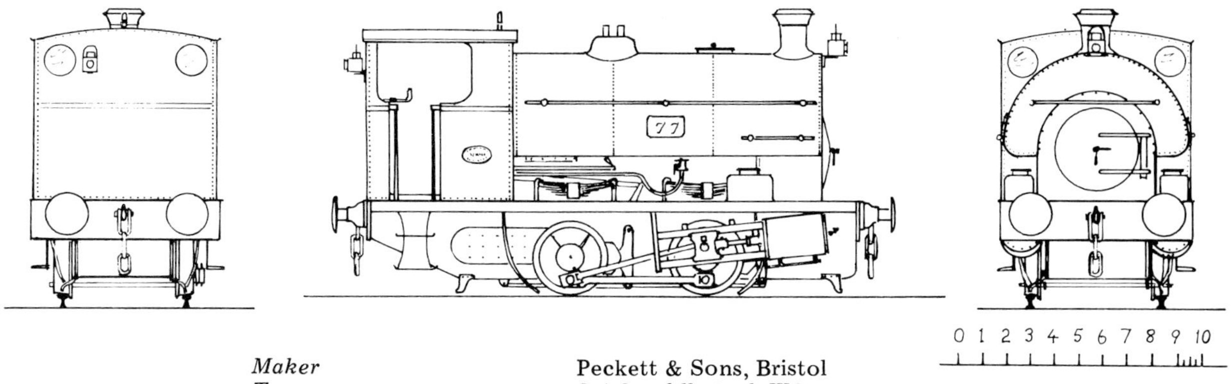

Maker	Peckett & Sons, Bristol
Type	0-4-0 saddle tank W6
Other engines in same class	None
Length over buffer beams	20ft. 2in.
Length overall	23ft.
Maximum width	8ft. 2½in. (over footsteps)
Height over chimney	11ft. 10in. (approx.)
Cylinders	14in. x 22in.
Wheel diameters	3ft. 3in.
Wheelbase	5ft. 6in.
Weight (empty)	23tons 10cwt.
Weight (in working order)	29tons 0cwt.
Capacity (coal)	10cwt.
Capacity (water)	920 galls.
Boiler pressure	160 p.s.i.
Heating surface (total)	603 sq. ft.
Heating surface (tubes)	540 sq. ft.
Heating surface (firebox)	63 sq. ft.
Grate area	9.25 sq. ft.
Tractive effort	

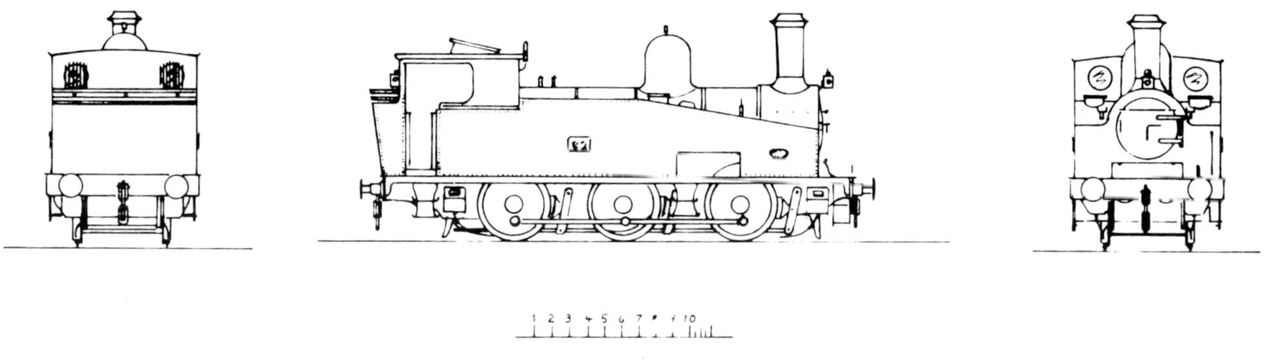

Maker	Kitson & Co., Leeds
Type	0-6-0 side tank
Other engines in same class	Total 3: Locos. 80, 81 & 82
Length over buffer beams	26ft. 6in.
Length overall	30ft. 6in.
Maximum width	8ft. 9in.
Height over chimney	13ft. 0in.
Cylinders	18in. x 26in.
Wheel diameters	4ft. 6in.
Wheelbase	13ft. 0in. (L to D 6ft. 9in.) (D to T 6ft. 3in.)
Weight (empty)	
Weight (in working order)	46 tons (approx.)
Capacity (coal)	2tons 5cwt.
Capacity (water)	1,350 galls
Boiler pressure	180 p.s.i.
Heating surface (total)	947.13sq. ft.
Heating surface (tubes)	864.29 sq. ft.
Heating surface (firebox)	82.84 sq. ft.
Grate area	18.7 sq. ft.
Tractive effort	21,060lb.

No. 79, a Hudswell Clarke long tank engine was delivered in June 1927, this engine being the last of the class and also the last new engine to be built at the Railway Foundry for the Manchester Ship Canal Company for the next twenty seven years.

The last locomotive to be ordered in 1927 was another Kitson. It was numbered 80 and was delivered in April. Although similar to the previous Kitson engines in most respects, the new engine sported a rear bunker. It was also fitted with side bunkers built into the tanks in the same manner as the previous engines. This gave it an increased coal capacity of 2¼ tons which was the largest amount of coal carried by any Kitson locomotive, as the next and final two were only fitted with a bunker in one tank as well as the rear bunker. The final two Kitson locomotives, Nos. 81 and 82, were delivered in 1929 on 12th and 17th January respectively.

Resident Engineer's Locomotive

Towards the end of 1931 the ex-contract engine, No. 6 *Longsight*, the Resident Engineer's (Partington) locomotive was sold, being replaced early the following year by a Hunslet engine, No. 83. This was a beautifully-proportioned 0-6-0 saddle tank with 12in. x 18in. inside cylinders. Like many of the contractor's engines which had worked in muddy conditions, No. 83 was fitted with wheel washers as the engine was intended to work on the tips at Statham. Of the earlier locomotives only Nos. 49 and 51 were fitted with wheel washers. To differentiate between the new locomotive and those used on traffic, the Resident Engineer specified an all over maroon livery which, when new, looked very tasteful but when dirty looked rather drab.

Mechanical Engineer's Locomotive

No. 84, the sister engine to No. 83, was delivered in May 1931, for use by the Mechanical Engineer's Department and was stationed at Mode Wheel shops. They were a pleasing class of engine in running, maintenance, and appearance, No. 84, particularly being kept highly polished for the whole of its life. With the arrival of No. 84, another ex-contractor's locomotive, No. 2 *Romilly*, was sold.

Locomotive	*Maker*	*Arrived*	*First Duty*
83	Hunslet Engine Co.	13.1.1932	
84	Hunslet Engine Co.	13.5.1932	Mode Wheel Shed

During 1938, the Company disposed of the oldest locomotive of the fleet, No. 34 *Copenhagen*. It was not sold for scrap but as a working engine to the Lancashire Steel Corporation. Also in the same year No. 26 *Bristol* was sold to W. I. Arnott Young & Company and although it went as scrap it is not recorded whether it was actually put to the torch. Both locomotives were working at Runcorn at the time of disposal and were replaced by sister saddle tank engines Nos. 75 and 78. There they were to stay until the railway at

When locomotive No. 80 arrived on the Canal railway system, the Kitsons were already old favourites but the addition of the rear coal bunker on this, the last class of Kitson to be introduced to the Manchester Ship Canal, made it even more welcome than the previous class. Built in 1927 (Works No. 5321), the engine served most of its time at Partington and Ellesmere Port.

Courtesy A. Appleton

This particular photograph of locomotive No. 84, taken on 18th May 1954 in Mode Wheel yard, shows the engine in what appears to be a black livery, but in reality the colours beneath the many years of daily-applied coatings of colza oil was dark blue (more slate grey), edged with black and lined in white. It is interesting to note that many serious students of the Manchester Ship Canal Railway believe that the Company had its engines painted black but this was not so.

A. Appleton

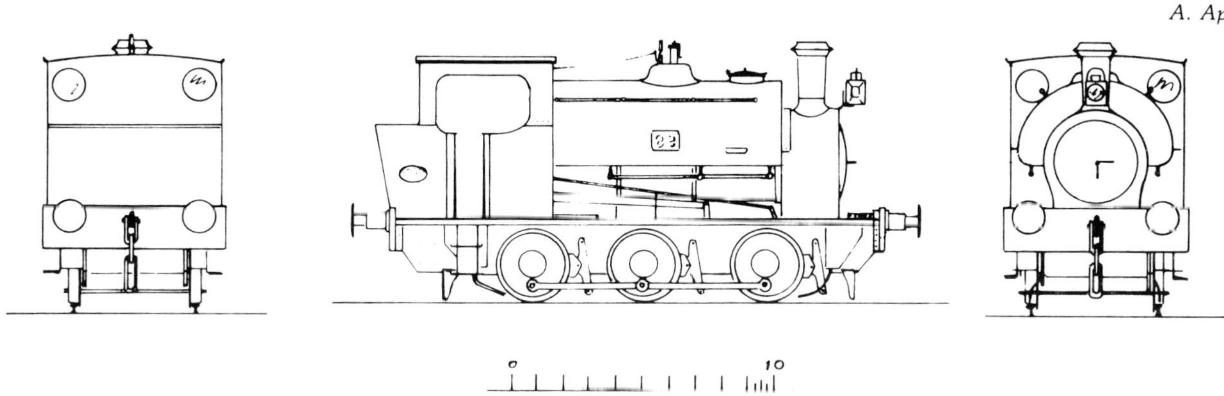

Maker	Hunslet Engine Co., Leeds
Type	0-6-0 saddle tank (Code 'Stman')
Other engines in same class	Total 2: Loco. 83 & Loco. 84
Length over buffer beams	19ft. 11¼in.
Length overall	23ft. 5in.
Maximum width	7ft. 6½in.
Height over chimney	10ft. 9in.
Cylinders	12in. x 18in. (inside)
Wheel diameters	3ft. 0in.
Wheelbase	9ft. 6in.
Weight (empty)	19 tons 0 cwt.
Weight (in working order)	24 tons 10 cwt.
Capacity (coal)	20 cwt.
Capacity (water)	550 galls.
Boiler pressure	160 p.s.i.
Heating surface (total)	422 sq. ft.
Heating surface (tubes)	376 sq. ft.
Heating surface (firebox)	46 sq. ft.
Grate area	7.8 sq. ft.
Tractive effort	8,640 lb. at 75% boiler pressure

Runcorn Docks finally closed down in 1952. Although the disposal of No. 26 and No. 34 reduced the fleet strength, the Company could evidently spare them as there were no moves to buy replacements, at least not immediately. It is not clear whether the Mechanical Engineer intended to purchase further locomotives at the time but, with the advent of World War II a heavy burden was placed on the railway to deal with the increased traffic coming from firms that had turned over to the production of strategic materials and munitions. In the event, the Manchester Ship Canal Railway, like the rest of the country's railways, suddenly found themselves in need of all the motive power they could lay their hands on. Obviously locomotive manufacturers were not in a position to supply engines, for they too had been put to work on armaments and Government work. Therefore, the answer was to buy second-hand. One such engine was *Jeannie*, a Hudswell Clarke short tank engine of the 'Canal' class.

When the Manchester Ship Canal bought *Jeannie*, she was taken into Mode Wheel shops for overhaul, which included a change of boiler and new wheels. The engine went into shops on 18th July 1940 and came out on 21st September with the letters MSC and the number 34 painted on its tanks.

During the war period, docks and railways came under the direction of the Ministry of War Transport Department and the visible evidence of this manifested itself in the movement of locomotives. The Manchester Ship Canal sent two engines to work for Lever Bros. at Port Sunlight on the Wirral. No. 15, a 'Jazzer', was put to work at Bromborough Dock and No. 29 worked at the main Sunlight factory. The dates during which the locomotives were away from the Manchester Ship Canal are not recorded, but No. 15 was given a light overhaul at the end of 1942 and a heavy overhaul betweeen 8th August and 3rd October 1944 at Mode Wheel shops, therefore, it would appear that the engine was away from Manchester during that period. No. 29 however, was sent to Port Sunlight after a light overhaul at Ellesmere Port, where it had been working. At the completion of the duty it was returned to Mode Wheel shops for a heavy overhaul. The date that No. 29 entered shops was 24th June and it was out-shopped on 30th August 1944.

Nevertheless, the transfer of locomotives was not all one way, for the Manchester Ship Canal was also allocated extra motive power, and this was made up of eight engines from the Port of London Authority and about the same number from the United States Army Transportation Corps. In the case of the Port of London Authority engines, they had been sent to various 'safe' ports because the London group of docks was receiving such a hammering from bombing that they were almost unworkable. The first batch to arrive at Manchester included Nos. PLA39, PLA40, PLA42 and PLA49. They were received on 24th September 1940. A short time later, Nos. PLA50, PLA51, PLA52 and PLA63 arrived but there is no record of the date on which they were received. Details of their movements are scant, Nos. PLA42 and PLA50 with two others working at Ellesmere Port, whilst, Nos. PLA51 and PLA52 were stationed, for some time, at Partington. Nos. PLA39, PLA40 and PLA63 appear to have worked most of their time at Salford. They gave sterling service at a time when help was most needed, for the Manchester Ship Canal Railway, during the war period, was at full stretch every hour of the day.

United States Army Locomotives

The year 1942 saw the arrival of the locomotives of the United States Army Transportation Corps. Unlike the Port of London Authority engines which were manned by Manchester Ship Canal staff, the American engines arrived complete with US Army crews. There were four engines at Mode Wheel shed and, for a while, two were stationed at Ellesmere Port. The locomotives were a standard type of 0-6-0 side tank which was built in large quantities for the US Government by several US locomotive builders including Vulcan Ironworks, Baldwins, Alco and others. They were unmistakably American with their bar frames, clamped smokebox door and outside cylinders, which seemed to be hanging in fresh air! The sand domes, situated front and rear of the steam dome, were a mixed blessing for although the sand remained dry, the job of refilling the containers was awkward. Two of the many good points in the locomotive's favour were the rocker bars in the firebox, and drilled firebox stays which showed a whisp of steam if that particular stay was fractured. A rather dubious feature was the cotton waste which lay in the axlebox keep, serving as an oil pad.

As the war in Europe progressed, the engines were shipped out, mostly to France. The exact date at which the last US Army engine left the Canal is not known but No. 4326, later Southern Railways No. 74, was still at Salford at the end of November 1944. A few examples of the class are still to be seen, with ex-SR No. 72 on the Keighley & Worth Valley Railway, and others on the Kent & East Sussex Railway at Tenterden and on the Bluebell Railway at Sheffield Park. After the war had ended, moves were made to return some of the Port of London Authority locomotives to London. Before they were despatched, a thorough examination was carried out, with special attention being paid to the tyres and axle bearings, for they were to be returned to London by rail. With the exception of No. 63, which had been completely overhauled in April/May 1944, all locomotives were overhauled.

Post World War II Locomotives

The next locomotives to be purchased by the Manchester Ship Canal were ex-War Department 0-6-0 saddle tanks, the 'Austerities'. This class of engine was the British equivalent to the US Army locomotives and was built in large numbers by

List of PLA Locomotives Working on the MSC Railway during World War II

No.	*Maker and Works No.*	*Year*	*Wheel Arrangements*	*Cylinders*
39	Andrew Barclay	1911	0-6-0ST	OC 16″ x 18″
40	Andrew Barclay 1238	1911	0-6-0ST	OC 16″x 18″
42	Andrew Barclay 1300	1912	0-6-0ST	OC 16″ x 18″
49	Hudswell Clarke & Co.	1915	0-6-0ST	OC 16″ x 18″
50	Hudswell Clarke & Co. 1153	1915	0-6-0ST	OC 16″ x 18″
51	Hudswell Clarke & Co. 1154	1915	0-6-0ST	OC 16″ x 18″
52	Hudswell Clarke & Co. 1155	1915	0-6-0ST	OC 16″ x 18″
63	Hudswell Clarke & Co. 1324	1918	0-6-0ST	OC 16″ x 18″

World War II design ex-WD 'Austerity' No. 75304, built in 1945 by the Vulcan Engine Co. Newton-le-Willows (Works No. 5294). No. 85 had only just emerged from the paint shop at Mode Wheel with the new MSC livery after being completely overhauled. The engine is seen outside Mode Wheel shops in the early months of 1948, and standing alongside is Jack Threadgold, Locomotive Superintendent and Alf Blanchard, Locomotive Foreman (nearest the buffer beam). Note the original two tread footsteps which were later replaced by three tread versions. All three 'Austerities' were dealt with in the same way.

Courtesy Mrs Loddington

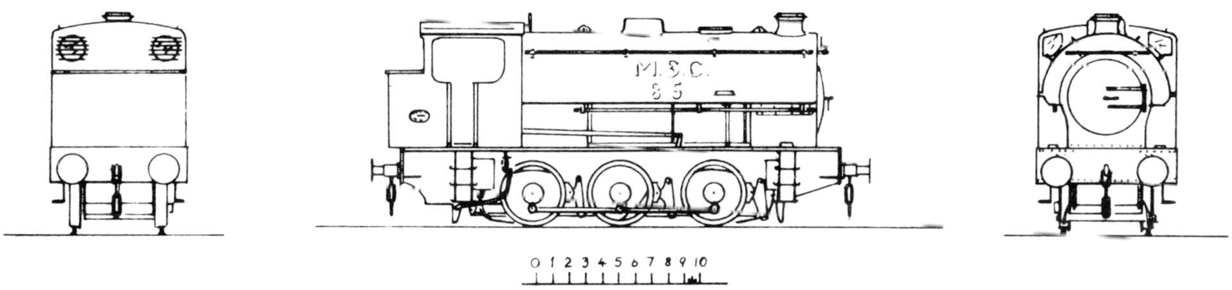

Maker	Vulcan Engine Co., Hudswell Clarke & Co., Hunslet Engine Co.
Type	0-6-0 saddle tank Ex-WD ('Austerity')
Other engines in same class	Total 3: Locos. 85, 86 & 87
Length over buffer beams	
Length overall	30ft. 4in.
Maximum width	
Height over chimney	12ft. 1½in.
Cylinders	18in. x 26in. (inside)
Wheel diameters	4ft. 3in.
Wheelbase	11ft. 0in.
Weight (empty)	37 tons 11 cwt.
Weight (in working order)	48 tons 3 cwt.
Capacity (coal)	2 tons 5 cwt.
Capacity (water)	1,200 galls. (imperial)
Boiler pressure	170 p.s.i.
Heating surface (total)	960 sq. ft.
Heating surface (tubes)	872.5 sq. ft.
Heating surface (firebox)	87.5 sq. ft.
Grate area	16.82 sq. ft.
Tractive effort	21,060lb. at 75% boiler pressure

several locomotive builders to a design by the Hunslet Engine Company. The earlier Manchester Ship Canal engines, Nos. 83 and 84, show a certain similarity, especially around the cab and bunker. The first one to be delivered came in late 1947 and was given the Manchester Ship Canal number 85. Originally War Department No. 75304, it was built in 1945 by the Vulcan Foundry, Newton-le-Willows. It was followed soon after by ex-War Department No. 71493, built by Hudswell Clarke & Company in 1944, and was number MSC 86. The engine went into Mode Wheel shops on 30th October 1947 and was brought out on 12th December. Ex-War Department No. 71441, built in 1945 by Hunslet, arrived at Salford and entered Mode Wheel shops on 25th May 1948. After the usual light overhaul the engine, by then MSC 87, left shops on 26th June. Apart from being overhauled, all three engines were fitted with a raised wooden floor in the cab, the former steel plates being unacceptable to the Manchester Ship Canal men, and small wonder, for a steel floor is terrible to work on in winter. Also the two tread footsteps were modified to three tread. The most visual changes were in the painting which followed the revised Manchester Ship Canal colour scheme of light grey with black and white lining. The letters 'MSC' and the number were painted on each side of the saddle tank in white with black shading.

The 'Austerities' were grand engines and a tribute to the designer. They were easy to work and extremely reliable, and they were very pleasing to the eye. As the term 'austerity' suggests, the engines were designed to cut out all unnecessary work and equipment in order to save time and materials. A major saving of both was the extensive use of welding in their construction.

Up until the time when the 'Austerities' arrived, the heaviest engines were the Kitsons at 44 tons 7 cwt, but the new engines became the new heavyweights, weighing in at 48 tons 4 cwt in working order. Surprisingly, the tractive effort of the 'Austerities' at 75 per cent boiler pressure, worked out exactly the same at 21,000lb. as the Kitsons, yet the boiler pressure was, at 170p.s.i., 10p.s.i. less than that of the earlier engines, and the grate area was less than that of the Kitsons. However, the overall heating surface of the 'Austerity' boilers was more than that of the Kitsons.

A feature, which deserves mention, was the tapered 'birdcage' spark arrester which was certainly effective, yet less likely to impair the efficiency of the boiler than that which was fitted to all earlier Manchester Ship Canal locomotives. Another feature was the gently-modulated tone of the tubular whistle, or rather, hooter. Springing was of the underslung type with extremely simple links and brackets, having no adjustment whatsoever. Even so, broken springs were no more prevalent than on other engines with more complicated spring tension adjustment systems. The engines were a great success, particularly at Partington and Ellesmere Port, where they finally worked.

Hudswell Clarke 0-6-0 Saddle Tank Locomotives Nos. 88, 89 & 90

With the arrival of any new type of locomotive, there was always a considerable stir of interest, but when No. 88 was shunted into the yard at Mode Wheel shops, there was more than just interest, for surely there had never been an industrial steam locomotive designed which was so handsomely proportioned. In fact, many enthusiasts have voiced the opinion that No. 88 and its sister engines, Nos. 89 and 90, were amongst the most magnificent industrial engines ever built. The livery was the Manchester Ship Canal light grey, lined in black and white.

Designed and constructed by Hudswell Clarke & Company, these 0-6-0 saddle tank locomotives, weighing 52½ tons in working order, were the heaviest ever to be used by the Manchester Ship Canal Railway Department. They were, of course, a train running engine, and like the Kitsons and 'Austerities' worked most of their years at Partington or Ellesmere Port.
However, No. 90 worked for a considerable part of its first years at Salford. The wheelbase was 12ft. and all wheels were flanged which restricted the locomotive to sections which included curves of no less than 230ft. radius. They were wider than any previous engine, the measurement over the outside cylinders being 8ft. 7¾in.

During the period in which the engines were being built, the Manchester Ship Canal men heard various rumours, which suggested that they were to be fitted with superheaters, Walschaerts valve gear, and electric lights powered by their own generators. In fact, there were none of these refinements, and the valve gear was the usual Stephenson's link motion. One departure from the usual Manchester Ship Canal practice was the fitting of a mechanical lubricator which served the cylinders and valves, and which replaced the sight feed steam-operated lubricator fitted to all previous engines. The fitting of white metal spring-loaded glands, to the piston rods and valve spindles, was also an advance on the earlier square section asbestos gland packing. They were heavy engines to work especially for the fireman, although the grate area was, at 17.88ft., less than that of the Kitsons, and the coal and water capacities were identical to those of the Kitsons. They were fine engines and the Manchester Ship Canal men were very proud of them but, unfortunately, they had design faults on the axleboxes and there was also a certain amount of trouble with fireboxes. No. 90 was particularly troubled by hot bearings, the axle brasses being taken out and 'backed off' on several occasions, the brasses having a tendency to close in.

Locomotive	*Maker*	*Ex-Works*	*First Duty*
88	Hudswell Clarke & Co.	10th Feb. 1954	
89	Hudswell Clarke & Co.	1st Mar. 1954	
90	Hudswell Clarke & Co.	7th Apr. 1954	Salford

There were two faults in the design of the axleboxes, the first being the method of supplying lubrication. The system comprised three oil boxes, each with two trimmings, which fed all six wheels. From the oil boxes a pipe was led to a flexible pipe which, in turn, was connected to a union screwed into one side of the axlebox. It seemed obvious, at the time, that one trimming was inadequate, especially for a locomotive which would be used for heavy train running, and so it proved. The second design fault was in the method of fitting the axle brasses. Instead of having a strengthening flat section on the top, which would bed into a mating recess in the axlebox, they were semi-circular on the outside, the boxes being bored to accommodate them. Set screws were then fitted to hold the brasses in position. It could be argued that, with a more adequate supply of oil, the bearings would not have tended to close in on the journal, but even so, a more substantial bearing would not have closed in so easily, therefore not getting quite so hot. However, eventually it was decided to double the oil feed, and this was done by fitting a second piped union in the axlebox which fed the outer end of the bearing oil groove. A set of oil boxes was fabricated with extra trimming tubes fitted, which had double the oil capacity

A very unromantic view of the first engine of the final batch of steam locomotives ordered by the Manchester Ship Canal. The photograph, taken in the yard of the makers, Hudswell Clarke & Co., Leeds, during the early part of 1954, shows clearly the lines of the engine.

M S C Co.

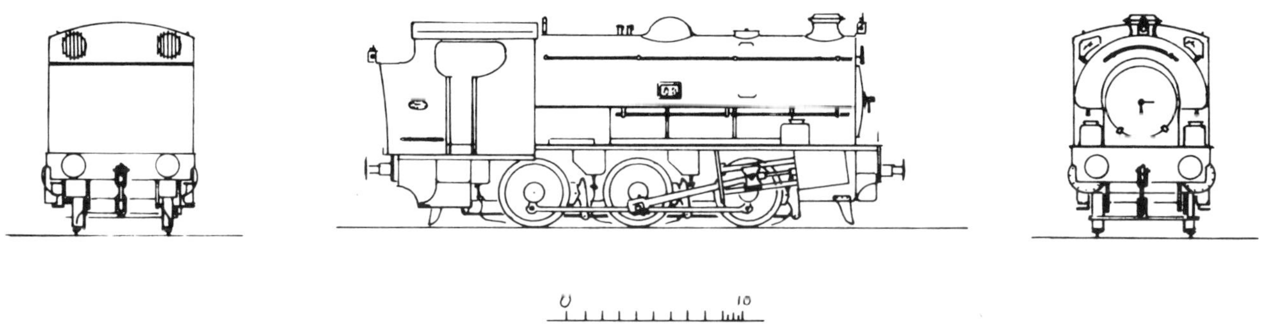

Maker	Hudswell Clarke & Co., Leeds
Type	0-6-0 saddle tank 88—90 class
Other engines in same class	Total 3: Locos. 88, 89 & 90
Length over buffer beams	28ft. 0in.
Length overall	31ft. 4in.
Maximum width	8ft. 7¾in.
Height over chimney	12ft. 6½in.
Cylinders	18in. x 26in. (outside)
Wheel diameters	4ft. 3½in.
Wheelbase	12ft. 0in.
Weight (empty)	42tons 0cwt.
Weight (in working order)	52tons 10cwt.
Capacity (coal)	2tons 5cwt.
Capacity (water)	1,350 galls.
Boiler pressure	180 p.s.i.
Heating surface (total)	1,039.03 sq. ft.
Heating surface (tubes)	945.94 sq. ft.
Heating surface (firebox)	93.09 sq. ft.
Grate area	17.88 sq. ft.
Tractive effort	Haulage power 797 tons (including loco) on level

A maker's picture of Hudswell Clarke diesel-mechanical locomotive *Enterprise*. This particular engine was tried out on the Manchester Ship Canal system and resulted in orders for fourteen diesel-mechanical locomotives of 204hp and one of 106hp.

Courtesy R. N. Redman

When diesel-electric locomotives Nos. 4001 *Alnwick Castle* and 4002 *Arundel Castle* arrived at Mode Wheel in 1958, many of the die-hard steam men prophesied a short life for the diesels, saying that they were not built like the steamers. But they were wrong on both points as the diesel-electrics have lasted, for they are built like steamers. The same sturdy Hudswell Clarke frames and undergear will last for many years, if they are allowed to remain in existence. The outward appearance, however, owes much to the designs of the Manchester Ship Canal Mechanical Engineer's Department, and the overall result is a fine pair of locomotives with a better than average appearance.

D. Thorpe

of the previous set. The trailing wheels, by the way, were lubricated by four trimming in a single box, which was located under a trap in the cab floor. As each locomotive went into Mode Wheel shops, the modifications were carried out, with the result that very little trouble was experienced from then onwards when the engines were working on normal traffic duties.

Even so, after the modifications had been effected, No. 89, whilst being moved between Latchford and Ellesmere Port by British Railways, suffered considerable damage to both trailing axle brasses and journals, when the lubrication did not prove sufficient whilst she was being towed at slow goods speed. The term 'slow goods' would appear to be rather suspect according to one eyewitness.

Dieselization

Even as the last steam locomotives were settling down to work, a locomotive trial was taking place at Salford, which was the first sign of the end of steam on the Manchester Ship Canal Railway. In the late summer of 1954, the Company borrowed a diesel locomotive for evaluation trials. This was an LMS type 350hp shunter weighing 55½ tons in working order. After a week of trials, during which the locomotive was based at Mode Wheel shops, the Engineers and Traffic departments found that a locomotive of this size was not powerful enough for all the duties to be undertaken. Actually, the locomotive developed trouble during trials and had to be towed away. A second diesel locomotive trial was arranged, at a later date, using a diesel-mechanical machine loaned from its builder, Hudswell Clarke & Company. *Enterprise*, as the locomotive was named, was used by Hudswell Clarke to demonstrate the particular type of unit to prospective customers. It is a wonder that the Manchester Ship Canal ever employed diesels, for the *Enterprise* also broke down, with fluid coupling trouble, and suffered the same ignoble fate as the LMS type locomotive, having to be towed away by a steam locomotive. It was five years before the results of the trials manifested themselves to the eager eyes of the Manchester Ship Canal railwaymen. On 17th March 1959, the diesel-electric locomotive *Alnwick Castle* arrived on the Canal and was taken to Mode Wheel shops where it was handed over to the Mechanical Engineer's Department for trials. The number given to the new locomotive was 4001 which was, of course, a complete change round from the numbering system then operating on steam locomotives.

Whilst some thought that the number 91 would have been more practical, most people seemed to accept the new numbering system, but were mystified by the choice of such a seemingly unconnected set of numbers, until they learned that the locomotives were designated by their horsepower rating followed by the class number. Actually the b.h.p. of the engine was 430 and not 400.

No. 4001 was an impressive unit with a very distinctive-shaped engine housing. Unlike most diesel locomotives, which have been designed to look like a loaf of bread, the engine housing of the new locomotive was built high enough to clear the top of the engine, but was only as far back as was absolutely necessary, after which it descended to the level of the top of the fuel tank. This allowed a centre window to be placed in the cab front. The chimney suggested steam rather than diesel, which was not surprising, for the pattern was the same as that used on locomotives Nos. 88, 89 and 90.

A surprising feature about No. 4001 was the colour, for instead of being painted in the livery adopted by the Manchester Ship Canal since 1945, grey fully lined out in black and white, she was painted in Sheffield Green. This may have been a different shade to the earlier green of the steam locomotives, but it was a good colour and won the approval of most of the staff. The outside frames and spring gear were painted black, the buffer beams, outside cranks and coupling rods being red. On the buffer beams the locomotive number was painted in white with black shading. The letters MSC on the sides of the cab were also in white.

Power was supplied by a Brush generator driven by a diesel engine built by the old established, but late, firm of National

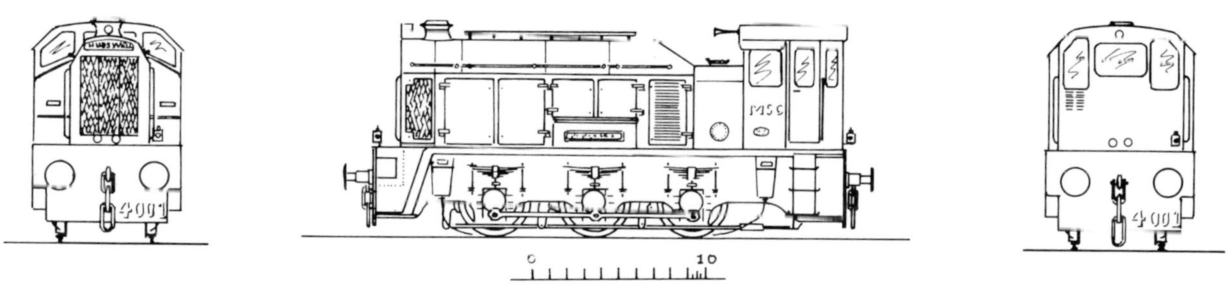

Maker	**Hudswell Clarke & Co., Leeds**
Type	0-6-0 diesel-electric
Locomotives in same class	Total 2: Nos. 4001 *Alnwick Castle* and No. 4002 *Arundel Castle*
Length over buffer beams	28ft. 0in.
Length overall	30ft. 10½in.
Width	8ft. 7in.
Height over chimney	12ft. 8½in.
Wheel diameters	4ft. 6in.
Wheelbase	11ft. 6in. (front to centre and centre to back, 5ft. 9in.)
Weight	51 tons 1 cwt.
Diesel power unit	National Gas & Oil Engine Co. 6 cyl. (normally aspirated)
Diesel power unit (rating)	430 b.h.p.
Diesel power unit (maximum r.p.m.)	750
Generator	Brush Traction Division (Hawker Siddeley)
Traction motors	Brush Traction Division (Hawker Siddeley)
Tractive effort	34,000lb.
Maximum speed	18 m.p.h.
Fuel tank capacity	
Test results	700 tons at 8m.p.h. on level
Test results	740 tons started on a 1 in 100 gradient

Gas & Oil Engine Company. The engine, a six cylinder in line unit delivered 430b.h.p. at 750r.p.m. It was similar in type to those installed in some large trawlers. Brush Traction motors were fitted to the leading and trailing wheels. The traction units were slung on the axles on two bearing journals, and were anchored to the locomotive frame through flexible support brackets. Between the motor pinion and the axle gear was an intermediate gear shaft, which could be jacked out of gear in the event of the locomotive having to be towed.

For safety reasons the control system was linked with a set of deadman's pedals. These were placed on either side of the control console. A seven second delay, on the power shut down and brake operation, gave the driver enough time to move from one side of the cab to the other where, on taking up his new position, he could reset the system by pressing down the alternative deadman's pedal. Bodily comforts were provided in the shape of hot water radiators fed from the engine cooling system, and an electric cooker, which was situated at the back of the cab. Sanding gear was operated by compressed air and was fitted to leading and trailing wheels only. A considerable amount of trouble was experienced for some time with the sanders but eventually most of the troubles were ironed out. At one period the makers of the sanding gear, Gresham & Craven, fitted liquid sanders, but they also gave trouble, chiefly due to the hardening of the mixture of sand and liquid. No. 4001 was fitted with this experimental system on 6th February 1960 and worked with it for several weeks before being refitted with the standard type of dry sander.

In April 1959, No. 4002 *Arundel Castle* arrived, and like No. 4001, was given acceptance trials. Both locomotives were troubled for months with small niggling faults such as failure of electro-pneumatic valves, blocked sanders, and faults on the engine shut down system etc., but when the teething troubles were finally sorted out, they settled down to become extremely valuable additions to the locomotive fleet, and were liked both by the footplate staff and the Traffic Department.

Diesel-Mechanical Locomotives

A second class of diesel made its appearance in December of the same year, this type being intended to replace the 'Jazzers' and the Hudswell Clarke short and long tank engines. After the diesel-electric locomotives, with their well-balanced, in fact, almost steam-engine appearance, the diesel-mechanical locomotive, No. 2001 *Ashdown*, was a very odd looking creature. This, of course, was only the impression gained by the Manchester Ship Canal people, for in the field of diesel locomotive design there are many oddities, just as there are in steam locomotives, which is brought about by the

The inauguration of the diesel fleet was marked by a ceremony outside Dock Office on 18th March 1959. The locomotive involved was Hudswell Clarke' diesel-electric No. 4001 *Alnwick Castle*. Present on that occasion was the Mayor of Salford, Alderman 'Sam' Davis, who was himself a Manchester Ship Canal engine driver.

Courtesy M S C Co.

Typical of the design of most industrial diesel locomotives, the Hudswell Clarke 204b.h.p diesel-mechanical locomotives are very functional, even ugly, but nevertheless they have proved to be a reliable, hard working, sturdily built unit. No. D14 (Works No. D1258), built in 1962, and the last of the class to be purchased, is shown at the head of a train in Salford yard. The white building in the background is Dock Office.

M S C Co.

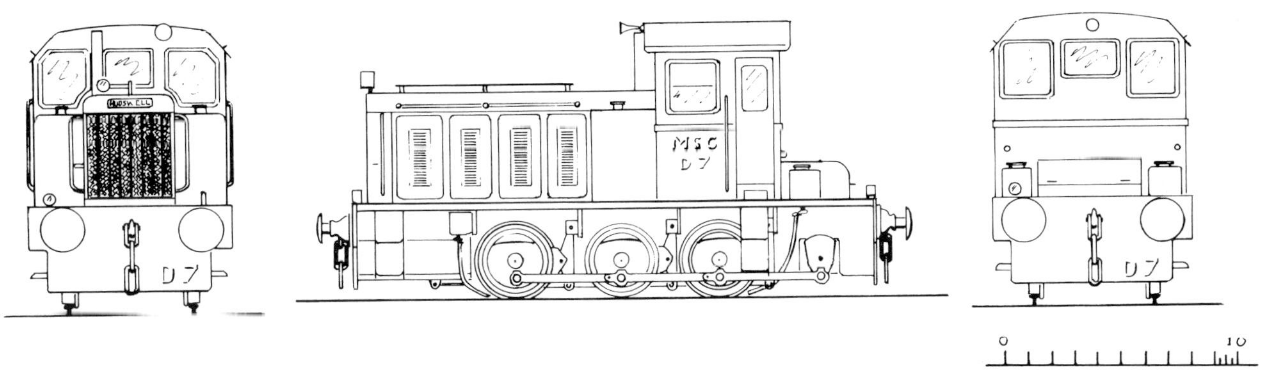

Maker	Hudswell Clarke & Co., Leeds
Type	0-6-0 diesel-mechanical
Locomotives in same class	Total 14: Nos. D1 to D14
Length over buffer beams	22ft. 2in.
Length overall	25ft. 4in.
Width	8ft. 8⅞in.
Height over cab	12ft. 0³/₈ in.
Wheel diameters	3ft. 4½in.
Wheelbase	8ft. 4in.
Weight (in working order)	36 tons
Diesel power unit	Gardner 8L3 8 cylinder in-line (normally aspirated)
Diesel power unit (rating)	204 b.h.p.
Diesel power unit (maximum r.p.m.)	1,200
Fluid coupling	Fluidrive Engineering Co. Ltd., Type Scr5 Size 23
Gearbox	Bostock & Bramley Ltd., Stalybridge. S.S.S. Power flow 3 speed F/R
Speeeds (m.p.h.)	1st gear 3.82: 2nd gear 6.73: 3rd gear 11.5
Tractive effort (lb)	1st gear 16,500: 2nd gear 9,400: 3rd gear 5,500
Fuel tank capacity	300 galls. (imperial) (left-hand tank 130 galls: right-hand tank 170 galls.)

placing of gearboxes, and the choice of power, transmission etc. One major difference in this respect, between No. 2001 and *Enterprise* was in the installation of the gearbox which was situated at the front end of *Enterprise* and at the rear end of No. 2001. By placing the gearbox at the rear, the cab had to be placed several feet forward, which was not a bad thing, for it gave the driver a slightly better view of the front end of the locomotive, a feature which is always desirable when working around dock areas.

Unfortunately, the placing of the gearbox so close to the cab made the working environment even noisier than that usually associated with diesel traction, and this was one of the main grumbles from footplate staff. Even so, the locomotives emerged as reliable and hard-working units, proving particularly useful on gradients. As with all previous locomotives from the Railway Foundry, the diesel-mechanicals are solidly-made machines, built in the usual methodical Hudswell Clarke manner.

Power is provided by a Gardner 8L3 diesel engine which develops 204b.h.p. at 1,200r.p.m. This is connected to the gearbox through a fluid coupling which can be made inoperative by the manipulation of a small lever in the cab. The SSS (Synchronised Self Shifting) gearbox drives the jack shaft which in turn is connected to the wheels by coupling rods. Brakes and sanders are air-operated. Unlike the diesel-electric locomotives, No. 2001 has no deadman's pedal or handle; in fact no automatic fail-safe devices in the event of driver collapse. Admittedly, the top speed was only around 8m.p.h. but in view of the crowded areas in which the Manchester Ship Canal Railway runs, it seems rather odd that no such devices were fitted.

The locomotive was relatively trouble free, the only major modification being the lengthening of the spring pillar brackets and the fitting of bearing pads between the axlebox and the spring pillars. Minor modifications included the replacement of the warm water radiators in the cab with car type blower heaters, and the cladding of the control cabinet with plywood panelling. The cabinet runs the full width of the cab and, being constructed of steel, was extremely cold to touch, even on mild days. For the preparation of food, a single burner Calor Gas cooker was installed on top of the cupboard at the back of the cab.

For some reason the second diesel-mechanical locomotive to arrive was numbered simply D2 and was unnamed. In fact, *Ashdown* was the last named engine to be purchased by the Company, and its number was later changed from 2001 to D1, in line with other locomotives of the class.

Resident Engineer's Locomotives

The diesel-mechanical locomotives were followed, in July of that year, by a similar type of locomotive but one which differed on three main points. First it was smaller, weighing 21 tons 6cwt against the 35½ tons of the 2001 class. Also it was powered by a Gardner six cylinder 6LH engine producing 106b.h.p, and finally, the gearbox had been placed at the forward part of the locomotive in the same manner as *Enterprise*.

The new locomotive was ordered for the Resident Engineer's Department to carry out the duties formerly done by locomotive No. 83. The new locomotive was painted maroon and given the number E1. The numbering system, at this point, was becoming more complicated!

Locomotive	*Maker*	*Arrived*	*First Duty*
4001	Hudswell Clarke & Co.	17.3.1959	Salford
4002	Hudswell Clarke & Co.	Apr. 1959	Salford
D2001 (D1)	Hudswell Clarke & Co.	Dec. 1959	Partington
D2	Hudswell Clarke & Co.	Jan. 1960	Partington
D3	Hudswell Clarke & Co	Feb. 1960	Partington
D4	Hudswell Clarke & Co.	Feb. 1960	Partington
D5	Hudswell Clarke & Co.	Mar. 1960	Partington
D6	Hudswell Clarke & Co.	Apr. 1960	Partington
E1	Hudswell Clarke & Co.	10.7.1960	

Diesel-mechanical locomotive No. E1, a Hudswell Clarke (Works No. D1199), built in 1960, is seen at Mode Wheel yard in 1980. With windows boarded up against attacks by vandals, and the paintwork in a battered condition, the more or less redundant locomotive looks a sorry sight indeed.

D. Thorpe

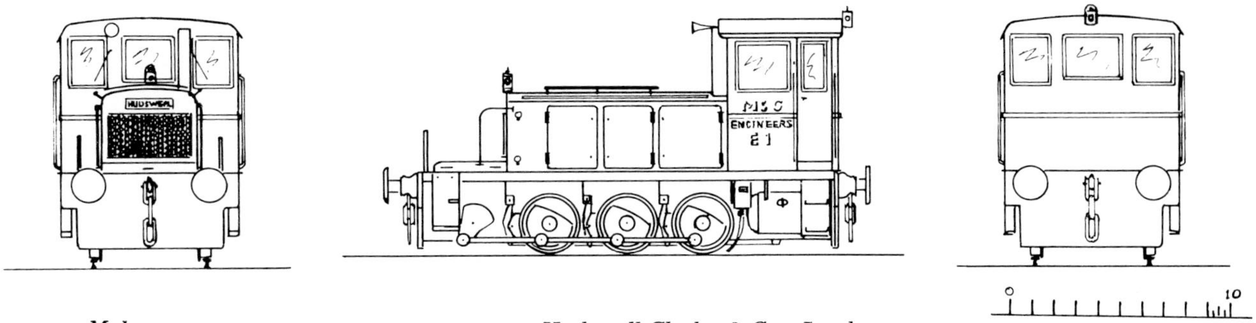

Maker	Hudswell Clarke & Co., Leeds
Type	0-6-0 diesel-mechanical (Engineer's locomotive)
Locomotives in same class	Total 1
Length over buffer beams	19ft. 2in.
Length overall	22ft. 4in.
Width	8ft. 2in.
Height over cab	10ft. 4½in.
Wheel diameters	2ft. 10¾in.
Wheelbase	7ft. 0in. (L to centre 3ft. 6in. and centre to T 3ft. 6in.)
Weight (in working order)	18 tons 2 cwt.
Diesel power unit	Gardner 6LW
Diesel power unit (rating)	106 b.h.p.
Diesel power unit (maximum r.p.m.)	
Fluid coupling	
Gearbox	
Speeds (m.p.h.)	
Tractive effort (lb)	
Fuel tank capacity	

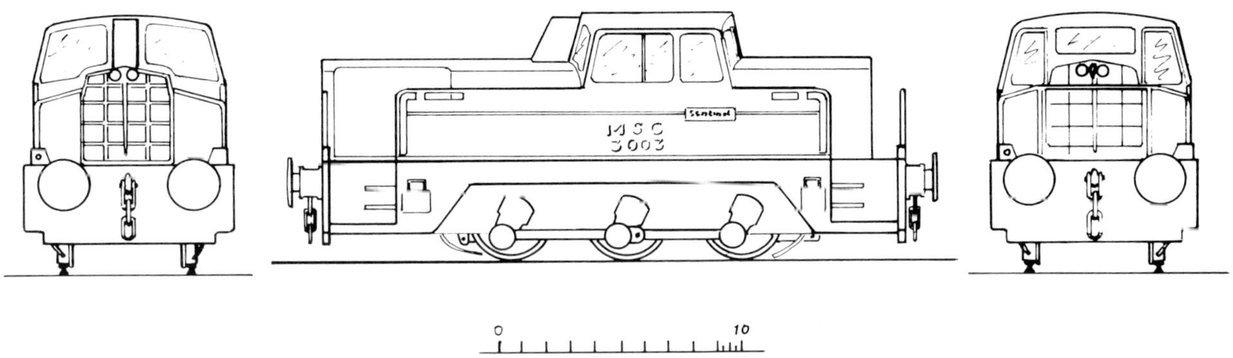

Maker	Rolls-Royce Sentinel Ltd., Shrewsbury
Type	0-6-0 diesel-hydraulic
Locomotives in same class	Total 5: Nos. 3001 to 3005
Length over buffer beams	23ft. 10½in.
Length overall	26ft. 10½in.
Width	8ft. 6in.
Height over cab	10ft. 6in.
Wheel diameters	3ft. 6in.
Wheelbase	9ft. 8in.
Weight (in working order)	48 tons (48.770 tonnes)
Diesel power unit	Rolls-Royce C8SFL with Rootes type supercharger
Diesel power unit (rating)	283 b.h.p.
Diesel power unit (maximum r.p.m.)	1,800
Torque converter	Rolls-Royce CF11500 (MS300 (stall torque ratio 11.64:1
Final drive to centre axle	Self Changing Gears Ltd., (forward/reverse)
Overall reduction	
Tractive effort	28,800 lb.
Maximum speed	17 m.p.h.
Fuel tank capacity	316 galls. (imperial)

During 1962, the second batch of 200hp type locomotives arrived, these being Nos. D7 and D8, which were delivered in February, whilst No. D14, the final engine of the class, arrived in late July.

Locomotive	*Maker*	*Arrived*
D7	Hudswell Clarke & Co.	Feb. 1962
D8	Hudswell Clarke & Co.	Feb. 1962
D9	Hudswell Clarke & Co.	Mar. 1962
D10	Hudswell Clarke & Co.	Mar. 1962
D11	Hudswell Clarke & Co.	Apr. 1962
D12	Hudswell Clarke & Co.	May 1962
D13	Hudswell Clarke & Co.	May 1962
D14	Hudswell Clarke & Co.	Aug. 1962

The largest number of the class worked at Salford with the bulk of the remainder being stationed at Partington. No. D7 went to Acton Grange, staying at this section for over two years before being sent back to Mode Wheel shops for major servicing. Livery was the same as that carried by Nos. 4001-4002.

Rolls-Royce Sentinel 0-6-0 Locomotives

Yet another set of numbers was added to the locomotive list when the first of the 300hp locomotives, No. 3001, arrived at Ellesmere Port in June 1963. If the diesel-mechanical locomotives were described as odd looking, the Rolls-Royce Sentinel diesel-hydraulic units could be called chic. An odd word to describe a work-horse but, there is no doubt about it, they were of very sleek design. With their car type engine housing and fuel tank plus, of course, the radiator grille with its symbolic chrome-plated sword, and chrome-plated headlamps, and also the angled surfaces of the upper part of the cab, they would not have looked out of place parked in front of a stately home, rather than in the industrial setting which was their home. The dark blue colour, with lining and lettering picked out in yellow, added to the rather classical appearance. The pity is, that many of the small fittings such as door locks, slides and window fitting were more suited to road vehicles than to locomotives. These were small faults but were responsible for many of the occasions when the locomotives had to be withdrawn from service, and it must be said, that this reflected badly on the Rolls-Royce image.

When diesel locomotives are discussed amongst railway enthusiasts, the terms 'loaves of bread' and 'boxes on wheels' are used, but this surely could not be so where Rolls-Royce Sentinels are concerned. This picture, of MSC Rolls-Royce Sentinel No. 3001 on the dockside at Ellesmere Port, clearly shows its superb outward appearance. Five locomotives of this type were purchased after trials were conducted at Ellesmere Port West between 23rd May and 27th June 1962, using a similar unit, but with a 311hp engine.

Rolls-Royce, Courtesy M S C Co.

Apart from these silly faults, the Rolls-Royce-powered 0-6-0 locomotives have proved fairly trouble-free. The diesel engine was a C8SFL unit coupled to a Rolls-Royce torque converter which, in turn, drove a self-changing gearbox mounted directly on the centre axle. The gearbox had a reduction ratio of 11.64 to 1 and incorporated a forward-reverse mechanism. Engine-road speed ratios were taken care of by the torque converter, and all wheels were coupled by side rods. A Rootes type supercharger took the engine up to a rated top r.p.m. of 1,800, very much more smartly than the previous diesels with their normally aspirated units. The feel of Nos. 3001-5 and the later 0-4-0 locomotives was rather akin to a sports car. At a weight of 48 tons in running order, it was quite a sports car, and certainly a top speed of 17m.p.h. was sporty enough for the complex railway system on which it worked.

By October 1963, the order for the 0-6-0 Sentinels had been completed, bringing the total of the class to 5, all of which went direct to Ellesmere Port and Stanlow where they more or less worked permanently.

Locomotive	*Maker*	*Delivered*
3001	Rolls-Royce Ltd.	19.6.1963 Ellesmere Port
3002	Railway Traction Department	27.6.1963 Ellesmere Port
3003	Shrewsbury	30.7.1963 Ellesmere Port
3004		6.9.1963 Ellesmere Port
3005		Oct. 1963 Ellesmere Port

Rolls-Royce Sentinel 0-4-0 Locomotives

In December 1963, an 0-4-0 version of the Sentinel was delivered. Yet again, the numbering differed, as this time the number was preceded by DH (diesel-hydraulic), but instead of starting at No. 1, as in the case of each of the previous classes, it commenced at No. 15 (DH15), the numbers actually following on from the diesel-mechanical class which ended at D14. To the tidy-minded, there was a niggling inconsistency about the numbering but, in fact, it did not create any difficulties, and from the workshop point of view, it was very helpful, especially when a particular type of locomotive had developed trouble away from the shops. A glance at the loco-

Locomotive No. DH20 (Works No. 10194), built in 1964, is pictured outside the Boiler Shop at Mode Wheel in 1965. This class of locomotive, with its chain drive to both axles, was rather a curiosity to the Manchester Ship Canal men when the first one arrived, but they quickly won new friends with their comfortable working conditions and ease of operation. One drawback was, that being of an 0-4-0 wheel arrangement, they suffered from the usual unstable ride as the steamers with the same wheelbase.

D. Thorpe

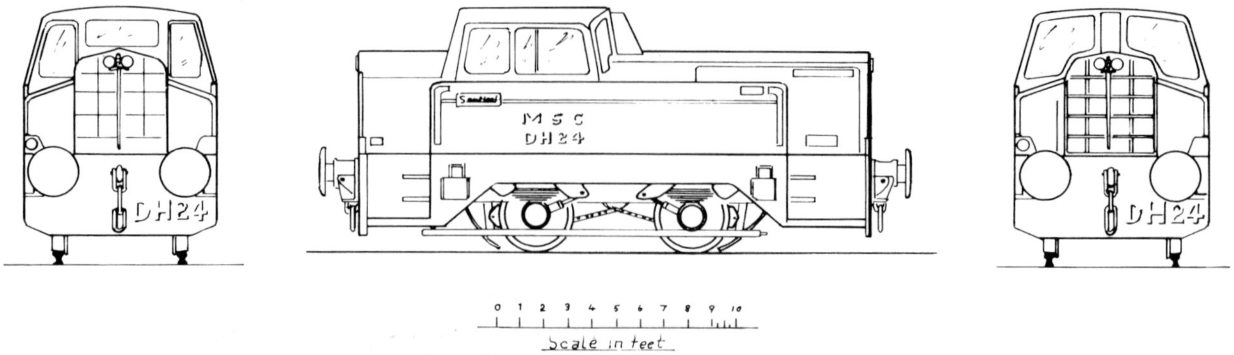

Maker	**Rolls-Royce Sentinel Ltd., Shrewsbury**
Type	**0-4-0 diesel-hydraulic**
Locomotives in same class	**Total 18: Nos. DH15 to DH32**
Length over buffer beams	**21ft. 3in.**
Length overall	**24ft. 4¾in.**
Width	**8ft. 0in.**
Height over cab	**10ft. 4in.**
Wheel diameters	**3ft. 2in.**
Wheelbase	**6ft. 6in.**
Weight (in working order)	**34tons (34.5 tonnes)**
Diesel power unit	**Rolls-Royce C6SFL with Rootes type supercharger**
Diesel power unit (rating)	**255 b.h.p.**
Diesel power unit (maximum r.p.m.)	**1,800**
Torque converter	**Rolls-Royce CF11500**
Final drive (to both axles)	**Self Changing Gears Ltd., (forward/reverse box to chains)**
Overall reduction	**10.45:1**
Tractive effort	**23,600lb**
Maximum speed	**17 m.p.h.**
Fuel tank capacity	**203 galls. (imperial)**

Driver Bernard 'Barney' Harcup at the controls of a Rolls-Royce Sentinel diesel-hydraulic unit. The cab, as can be seen, is of a clean, uncluttered appearance, graced with a well-designed console and conveniently placed controls.

Council of Industrial Design

In contrast to the control layout of the diesel units was the boiler front and controls of the earlier steam locomotives. The view is that of a Hudswell Clarke 'long tank'.

Courtesy W. Clayton

motive number would immediately register the class and would indicate what might possibly be the problem.

In many respects the 0-4-0 Sentinels were similar to the 0-6-0 version, but in drive they were decidedly not. In the tradition of earlier Sentinel steam locomotives, both the leading and trailing sets of wheels were chain-driven. The usual difficulty with chain-driven sprung axles was partially overcome by the use of an arrangement of swinging arms, which allowed the axles to move on an arc almost equal to the radius of the chain drive. Although chain-driven locomotives are not the average railwayman's idea of a 'proper' engine, they were received fairly well. Lubrication of the exposed chain was partially effected by an oil mist sprayer which worked intermittently, spraying directly on to the chain. With both leading and trailing axles being connected by the chains to a common output shaft at the gearbox, the coupling of the wheels, using side rods, was avoided. However, the system of swinging arms and suspension spring to constrain the wheel sets, produced problems in designing the braking equipment. Brakes were of the usual brake 'block-on-tyre' type but unlike the standard system of pulling all the block in one direction, it was judged necessary to have the brake hangers fitted at the rear of the trailing wheels, and at the front of the leading wheels. The application of the brake caused the blocks to move towards the centre of the locomotive. This obviously involved a rather complicated brake assembly and the resultant system, although effective, gave rise to a great deal of extra work for the workshop staff when brakes and chains were being adjusted. Brake gear is always a rough job with the adjustment screws being so vulnerable to attack by rust, but working on the brake gear of the 0-4-0 diesel-hydraulic class locomotives was not considered to be anything but hard work.

Apart from the same niggling faults which plagued the 0-6-0 Sentinels' door locks and slides, window catches and windscreen wipers, the diesel-hydraulics were well received by the locomotive crews. They were particularly happy with the smart modern style cab and the fingertip dual-control system. The contrasting styles of cab layout are shown on *page 142*. From the shunter's point of view, the platforms at the front and rear were a safer and more convenient place to ride whilst on shunting duty. The diesel-hydraulics at 34 tons operating weight, were the heaviest and most powerful 0-4-0 locomotives ever worked by the Manchester Ship Canal and were a natural choice to replace the Peckett locomotives at both Pomona and Ordsall.

A further seven 0-4-0 diesel-hydraulics were delivered in 1964 but, by that time, the total number of locomotives had fallen to 54, twenty four of which were steam locomotives. A good number of the steam locomotives were standing idle in the sheds, with boilers cold and brasswork dulled through neglect, but they were not to stand for much longer, for in 1964 a further batch was sold for scrap. There were 22 locomotives in all, and of these, three were 'Austerities' and three were the large Hudswell Clarke saddle tank engines, Nos. 88, 89 and 90, which were just over ten years old. It was sad to see these fine engines go to the torch, and quite a few of the

Both the Hunslet Engine Co. in 1900 and Thomas Hill Ltd. in 1966 are saying the same thing, that the Manchester Ship Canal Railway was important enough to be quoted as using their products.

Courtesy Stretford Library and Thomas Hill (Rotherham) Ltd.

MANCHESTER
SHIP CANAL COMPANY
CHOOSE

Sentinel for the arduous duties involved in dock shunting

"Sentinel" 325 b.h.p., 48-ton diesel hydraulic locomotive

ROLLS-ROYCE "SENTINEL" DIESEL HYDRAULIC LOCOMOTIVES AS OPERATED BY THE MANCHESTER SHIP CANAL COMPANY INCORPORATE ROLLS-ROYCE DIESEL ENGINES AND 3-STAGE TORQUE CONVERTERS. WITH A CONTINUOUS RATING FROM 1·15 M.P.H. STANDARD PRODUCTION MODELS SUPPLIED FROM 30 TO 75 TONS WEIGHT AND 230 TO 650 B.H.P.

Supplied by:—
Thomas Hill (ROTHERHAM) Ltd.
VANGUARD WORKS, KILNHURST, Nr. ROTHERHAM
Phone: Mexborough 2571

staff expressed their disappointment at the lack of success in selling them as working machines. It certainly is a pity that at least one specimen was not preserved, for they were the only ones of their type ever built. However, it is understandable that the Company, with such a costly railway programme to complete, would not be able to justify sentimental gestures of that nature.

In October 1965, the final phase of the dieselization programme commenced. Three 0-4-0 Sentinels were delivered, followed by another one in November, and the last half dozen were delivered in July 1966. In all, the Company then owned 40 diesels and 24 steam locomotives.

End of Steam

On Friday, 30th June 1966, the steam era came to an end on the Manchester Ship Canal Railway, when locomotive 67 returned to Mode Wheel shops at the end of her shift at

Locomotive	*Maker*	*Delivered*
DH15	Rolls-Royce Ltd.	Dec. 1963
DH16	Railway Traction Department,	Feb. 1964
DH17	Shrewsbury	Feb. 1964
DH18		Apr. 1964
DH19		Oct. 1964
DH20		Oct. 1964
DH21		Oct. 1964
DH22		Oct. 1964
DH23	Rolls-Royce Ltd.	Oct. 1965
DH24	Railway Traction Department,	Oct. 1965
DH25	Shrewsbury	Oct. 1965
DH26		Nov. 1965
DH27	Rolls-Royce Ltd.	Jul. 1966
DH28	Railway Traction Department,	Jul. 1966
DH29	Shrewsbury	Jul. 1966
DH30		Jul. 1966
DH31		Jul. 1966
DH32		Jul. 1966

Salford. There was no bunting or television cameras, as there had been for the arrival of No. 4001. The scene comprised a rather grimy engine and three men finishing a normal shift. However, on 6th July a ceremony was performed outside Mode Wheel shops to mark the passing of steam locomotives and the completion of the dieselization programme, and a tribute to the passing of steam also appeared in the Company's house magazine. Locomotive No. 67 did, however, have the luck to be spared and to work again, for in June 1967, a Mr Greenwood purchased her together with No. 31, a Hudswell Clarke short tank engine, for preservation. Also in 1967 the final batch of locomotives due for scrapping, was sold, and by the end of October only two steam locomotives remained. Nos. 32 and 70 were both laid up at the bottom of the shed at Mode Wheel shops, ostensibly as standby engines. They were brought out on several occasions, and the staff took care to keep them up to scratch, by polishing the brasses and wiping the outside down with oily rags. However, some say there were complaints from visitors to Weaste Cemetery about smoke nuisance, which led to the lapse of the insurance, and both locomotives were completely withdrawn from any form of standby duty. Like Nos. 31 and 67, they were fortunate in being bought for preservation.

Reduction in the Locomotive Fleet

By 1970, the situation, with regard to traffic, had reached such a state that even the diesel fleet was too large, and in November of that year two of the diesel-hydraulic Sentinels were sold to Bowater at Ellesmere Port. The year after, a further five of the same class were sold back to the Sentinel agents, Thomas Hill & Company. With the closure of the docks railway, in 1972, the Company sold off another seven of the same class, six going back to the agent and one being bought by British Salt. The reduced fleet then consisted of two Hudswell Clarke diesel-electrics, fourteen 204hp and one 106hp Hudswell Clarke diesel-mechanical, five 300hp 0-6-0 Sentinels, and four 230hp 0-4-0 Sentinels, one of which was being retained for spare parts. Diesel-hydraulic No. 22 was the cannibalized locomotive whilst diesel-hydraulics Nos. 23, 24 and 26 were fitted with tandem driving equipment at the rear

The changeover from steam to diesel was completed with a ceremony, held on the through route outside Mode Wheel shops, on 6th July 1966, with locomotives Nos. 31 and 32 representing the retiring steamers and diesel-hydraulic locomotives Nos. 27 to 32 (the last of the Rolls-Royce engines) representing the new order. Also present was the Mayor of Salford, Alderman F. Dewhurst, JP, the Managing Director of the Manchester Ship Canal Co., Mr W. G. Edmunds, CBE and many of the Company's staff.

Rolls-Royce Ltd., Courtesy M S C Co.

end, which allowed any two locomotives out of the three to be coupled together and worked by one driver. All three tandem-fitted locomotives were modified by the staff of Thomas Hill & Company at Mode Wheel shops during December 1971 and January 1972.

As the railway contracted and the tips at Statham were cut off from the remaining railway line, so the Resident (Civil) Engineer's Department found less use for their locomotive and subsequently No. E1 was taken to Mode Wheel shops for storage. For some time it was up for sale and I. C. I. were reported to have been interested in it, but nothing came of that and, some years after, No. E1 was still in store at Mode Wheel shops.

About the same time, four of the Hudswell Clarke diesel-mechanical locomotives were put up for sale and, in January 1976, Nos. D9 and D10 were purchased by the Hunslet Engine Company. Both engines were reported as being sold or hired to the NCB for work at Powell Duffryn, North Wales. The third locomotive, No. D4, was sold to the locomotive firm Andrew Barclay & Sons in February 1976, and was followed the sale of No. D1 (2001 *Ashdown*), to the Hunslet Engine Company in April. No. D4 was on loan to the Anglesey Aluminium Company to work at their plant near Holyhead whilst their own locomotive was being overhauled at the Hunslet Engine Company. The locomotive worked there for six months during the summer of 1976 before being returned to the Hunslet Engine Company.

At the close of 1983, the Manchester Ship Canal had 15 locomotives, comprising diesel-electrics Nos. 4001 and 4002, diesel-mechanicals Nos. E1, E2, D3, D6 and D11, and diesel-hydraulics Nos. 3001-5, and the tandem engines, Nos. DH23, DH24 and DH26.

Locomotive No. 67 is seen rolling past the sidings of Cranes Department at Salford yard on 30th June 1966. It is believed to have been the last steamer to be rostered for duty.

Courtesy H. Kenney

Locomotive No. 31 *Hamburg*, being winched on to a low loader in preparation for its removal to Haworth, central depot of the Keighley & Worth Valley Railway. The photograph was taken outside the Fitting Shop at Mode Wheel, in June 1967.

R. Johnson

Chapter Eight Rolling Stock

MSC Flat Wagon No. 444 was a purpose-built wagon which was used for the movement of locomotive boilers and tanks at Mode Wheel Workshops, and was purchased new from Ashbury Railway Wagon Co. in May 1898. The wagon, much rebuilt, lasted until 1973 when it was broken up with the withdrawal of all the Company's rolling stock. The plate was retained for preservation.

D. Thorpe

During the years between 1894 and 1972 the Company owned 6,720 wagons, at least 12 coaches and an unknown number of vans. Of the wagons used during the construction period, known to have been in the excess of 6,300, only 400 were retained by the Manchester Ship Canal for their own use after 1894. These were numbered MSC Nos. 1 to 340, 392 to 420, 446 to 457, 495, 546 to 549, 770 and 823. Quite a lot of the ex-contract wagons were of the unsprung bufferless type commonly used for shifting spoil, and most of these were taken by the Resident Engineer's Department.

Between July and August 1897 the Company took delivery of 51 wagons new from Ashburys, the same firm which had supplied so many of the wagons used in the construction. Further batches from the same firm were delivered over a period of two years, during which time a total of 181 wagons was purchased. A surprising number of these and some ex-contract wagons lasted until the end of the dock railway system in 1972, and one wagon, No. 375, which was the longest serving wagon on the railway, having finished its days as a weed-killing tank carrier for the Resident Engineers, was sold to Taylor Bros. on 12th March 1973. No. 444 was also an Ashbury wagon but was built for yard purposes at Mode Wheel shops. This particular wagon was an 18ft. long flat with a 9ft. wheelbase and was used to move boilers and tanks in and out of the shops. The wagon went to Taylor Bros., in February 1973 but the plate was saved and is in store for historical purposes. Obviously these wagons were much rebuilt and probably retained none of their original timber members. After November 1899 no further new wagons were purchased by the Company, all subsequent rolling stock being bought second-hand.

In the early years, the majority of box wagons were of 8 tons capacity, which reflected the power of the locomotives of that era. As larger locomotives were bought, the average load weight grew to 10 tons and later to 12 tons. There were, at that time, a few 15 ton bolsters in use. In later years the largest wagons were the ex-War Department 45 ton bogie bolsters, called 'Parrots' acquired from the Disposal Board in 1923 and 1927. The first batch, five in number, was given Manchester Ship Canal numbers 2252 to 2257 and cost £350 each. They were delivered on 7th June 1923. The next batch of 'Parrots', arrived on 22nd February 1927. There were twenty in this batch and they were numbered 2705 to 2724. As late as 1961 there were large bogie bolsters being purchased, the largest capacity being 40 tons, of which there were just five. These came from the Central Wagon Company and Bolton Wagon Company. Also purchased were five 30 ton bolsters, sixeen 20 ton bolsters, ('Macaws'), one 12 ton and one 10 ton bolster. The whole batch was bought for £7,465. Prices paid for wagons in the years before 1923 were not well recorded, but in 1923 the price for a second-hand 10 ton box wagon averaged about £25. By 1967, the year that the Manchester Ship Canal bought their last box wagon, the price for a similar wagon to those purchased in earlier years had risen to about £183. From 1945 until 1970, when the Company made their last rolling stock purchases of two brake vans, the amount paid out was over £536,000. Apart from the wagons used by the Resident Engineer's and Mechanical Engineer's Departments, all other Company wagons were used for internal traffic only.

A rather mysterious wagon was No. 3121A which was 'found' on the docks at Salford on 21st October 1953. How anybody could lose a wagon is an even bigger mystery!

The Company's Livery

Livery on the early Manchester Ship Canal stock was of two distinctive styles. One style and colour was adopted by the Resident Engineer's Department, whilst the Traffic wagons and those of the Mechanical Engineer's Department shared the same basic style, all over black with white lettering. The bold full lettered 'Manchester Ship Canal' was in two sizes of letters. The word 'Manchester' was picked out in 4in. x 1in. letters whilst the words 'Ship' and 'Canal' were picked out in 12in. x 2in. letters.

It will be seen that the wagon number painted on the side sheet is in 5in. x 1in. numerals and the wagon number on the frame, also the tare weight and capacity, are painted in 4in. x 1in. numerals.

Resident Engineer's Stock

Rolling stock belonging to the Resident Engineer's Department is normally painted an all over red oxide colour with the wording 'Manchester Ship Canal' along each side, and the wording 'Resident Engineer's Dept.' painted below.

In later years the time consuming task of painting the full wording was discontinued except in certain cases. The practice from about 1953 was to paint the letters MSC along the wagon sides in 18in. x 2in. letters as can be seen on the wagon illustrated on *page 152*. In all other respects, however, they were exactly as they had been previously. Lettering on bogie bolsters was similar to that of the early wagons except the words 'Manchester Ship Canal' were 6in x 1in. characters.

Brake Vans

Brake vans were not used in many sections of the system, in fact the only places where they were used were Stanlow, Acton Grange and Latchford. Actually they were not used at Acton Grange or Latchford to the same extent as they were at Stanlow, where they are still used regularly. The last mention of a brake for Latchford was on 21st January 1948, when a 20

An ex-contractor's wagon, No. 187, in use on the Company's Railway during the early years. The lettering on the sides was in the style first adopted by the Manchester Ship Canal. Note the dumb buffers which were typical of most of the contractor's rolling stock.

Courtesy Trafford Park Estates Ltd.

A medium sided 8 ton box wagon, MSC No. 454. This wagon was again an ex-contractor's vehicle and was without buffers. Many of these wagons were originally built by the Ashbury Railway Wagon Co. of Manchester.

Courtesy Trafford Park Estates Ltd.

Ex-War Department bogie bolster, MSC No. 2709, in use as a track carrier for the Resident Engineer's Department. The wagon was one of those bought from the Disposal Board in 1927 which were code-named 'Parrots'.

G. Holt and A. Wadsworth

A 10 ton box wagon, No. 3215, photographed in 1950. This wagon was one of a large batch purchased from The Rolling Stock Co., circa 1939. It is interesting to note that the Company's title is painted in the second style to be adopted by the Manchester Ship Canal Company.

G. Holt and A. Wadsworth

MSC brake van No. 6373 in the Stanlow Compound, along with 0-6-0 Rolls-Royce Sentinel No. 3001, in August 1978. Stanlow is the only section where brake vans are used. Nos. 6373 and 6372 were purchased from the Bolton Wagon Co. in August 1962, each for the sum of £260, both being sent direct to Ellesmere Port.

D. Thorpe

An Engineer's inspection coach behind Hudswell Clarke 1590, locomotive No. 79, built in 1927, in Mode Wheel yard, possibly in the early 1930s. The figure in the centre of the group is believed to be G. Leader Williams, Resident Engineer and son of E. Leader Williams the Engineer responsible for much of the planning of the Ship Canal before it was built.

Courtesy A. Appleton

Dock Labour Board coach No. 3 stands in a siding at Irwell Park on 2nd August 1966. This vehicle, ex-LMS, open third No. 7991, built in 1926 (first number 5682), was used to move dockworkers from Salford to the various outlying wharves when necessary.

I Higginbotham

Ex-GCR cashier's coach No. 1 is seen hauled by locomotive No. 28 *Belfast*, Hudswell Clarke No. 644 of 1903. The photograph was taken at No. 1 Grid, Salford on 3rd May 1947.

A. Appleton

ton van was delivered. It was given the Manchester Ship Canal number 4429, and sixteen years later, in September 1964, it was broken up but was not replaced.

Coaches

The number of coaches and vehicles used in that capacity was not well recorded, the first mention of a coach being on 21st November 1930, when Dock Traffic Coach No. 1 was delivered. The exact purpose of this coach is not clear but it is believed to have been used for the movement of the dockers from Salford to Irwell Park, in the same way as the later Dock Labour Board coaches. Following the purchase of the Dock Traffic Coach No. 1 came Railway Inspector's Coach No. 1. The user was to be the Resident Engineer, Mr George Leader Williams, youngest son of E. Leader Williams, the engineering brain behind the building of the Canal.

During September 1945, the wagon examiner reported Dock Labour Coach No. 1 as unfit for use, the result being that the wagon makers had to carry out a considerable amount of work on the coach before it was deemed fit for duty. There is very little record of Dock Labour Coach No. 1, the first No. 1 that is, for eventually there were two of them. It would appear that the first Dock Labour Coach No. 1 was probably the earlier recorded Dock Traffic Coach No. 1. The second coach with the title of Dock Labour Coach No. 1 was bought by the Company in 1946, but why it did not become No. 2 is not really clear. The new coach, which came from The Rolling Stock Company, Darlington, cost £220 and was delivered to Bridgewater Junction on 3rd April. It was a 50ft. ex-LMSR third class brake, No. 22209. This particular coach came to a sticky end when, on 3rd September 1960, locomotive No. 84, whilst putting some wagons into No. 5 road, 3 grid, forced the coach over the concrete stop block, leaving the bogies behind. A few weeks later during week ending 8th October, the coach was broken up by the Central Wagon Company.

Again records do not show when Dock Labour Coach No. 2 was bought, or even what type it was. However, it is known to have been an 8 wheeler similar to Dock Labour Coach No. 3. Dock Labour Coach No. 2 was recommended for scrapping on 28th October 1965 and this was done before the year was out. On 3rd June 1960, the final Dock Labour Board Coach, No. 3, was purchased from British Railways for the sum of £275.

Cashier's Coaches

Cashier's Coach No. 1 came into use early in 1946. As it was explained in Chapter 3, wages were paid out to staff along the Canal between Salford and Old Quay by launch, and on foggy days, by locomotive as far at Latchford and by hire car from Latchford to Old Quay. With the laying up, in 1945, of the launch *Dispatch*, which had been taking the cashiers along the Canal for many years, arrangements were made for the cashier, his assistant and a police escort to be carried to Latchford on the locomotive. It was not a satisfactory arrangement, especially from the fireman's point of view, as he had a job to fire the locomotive properly with such a crowd in the cab, and therefore it was decided to place a suitable vehicle at the cashier's disposal.

The records do not show when Cashier's Coach No. 1 arrived or who it was purchased from, and there is some difference of opinion as to what Company it originally belonged. One authoritative body of opinion stated that the coach had GN on the axleboxes, whilst a second, and equally informed, observer stated that it was an ex-GCR vehicle. After the coach had been through the shops, it emerged with the body painted the usual red oxide colour and a black frame. The 8in. x 1¼in. letters (approximately) MSC Co., were picked out on the narrow panel below the windows.

The second Cashier's Coach, No. 2, an ex-Midland Railway 4 wheeler, was delivered on 21st March 1953 and was ready for duty by 21st April. It was originally a brake, as the brakeman's 'ducket', protruding from the side, shows, and it bore the number M198718. The Central Wagon Company Ltd., were the suppliers. On arrival at Mode Wheel, where all the Cashier's Coaches have been stabled, the livery was changed to the same red oxide colour as No. 1. The Company's initials were, however, limited to MSC which were picked out in white letters approximately 4in. x 1in. high.

Like the previous cashier's coach, it was sparsely furnished with a fixed table and seats, and was heated by the usual cast-iron solid fuel stove. An interesting feature of the coach was the construction of the wheels which were a composite of wood and steel. They were made in this way in order that the rail and tyre noise would be lessened. Cashier's Coach No. 2 was broken up soon after Cashier's Coach No. 3 came into use.

The last coach, Cashier's Coach No. 3, was originally a horse-box! After it was delivered to Bridgewater Junction on 25th April 1965, it was put into shops for the most extensive series of alterations ever carried out on one of the Company's coaches. The coach entered the wagon shop on 29th April and was ready for service six months later after having, amongst other alterations, steel shutters which could be slid into position over windows and doors if the need arose. An alarm was fitted which could be connected to the locomotive so that the driver could give warning of any suspicious activity ahead. What happened to the unfortunate driver if anything criminal did occur is not clear. He would simply be locked out and left to fight it out! Lighting and heating was by butane gas, and a toilet was also installed. The vehicle was a wooden body mounted on a steel frame and was built at Earlstown in 1954, and numbered 1662. A 4 wheeler with oil bath axle-boxes, it was purchased from the Central Wagon Company Ltd., for the modest sum of £150, but the cost of converting it was several times that amount. It was a pity really that so much time, effort and money had gone into the conversion, for the cashier's use of the railway ended four and a half years later when, on 22nd April 1970, Cashier's Coach No. 3 made the last run. By that time the coach was only being taken as far as Partington, the line to Latchford being virtually closed after May 1967. The photograph shows the coach, windows boarded to repel vandals, at Mode Wheel on 12th October 1979. Usually the locomotive rostered for the Cashier's Coach was a 'Jazzer' but sometimes the Engineer's locomotive, No. 84, was used, and it was normally an 'invalid' driver's job.

Coaches and Vans for the Resident Engineer's Department

At least 246 wagons, vans and coaches have been used by the Resident Engineer's Department over the years. Apart from the spoil wagons and weed-killing tanks previously mentioned, there have also been water carriers, standard box wagons and bogie bolsters. A number of coaches and similar vehicles, purchased in the post-war years, were used as personnel and tool carriers, whilst others were designated for kitchen and mess room facilities. In the latter category were two coaches which deserve mention, even if it is only because of the names which they bore during their term of service on

Cashier's coach No. 2, standing in its usual 'parking' place outside the Boiler Shop at Mode Wheel yard. The photograph was taken about 1954, the coach at that time having been in use on the Manchester Ship Canal for about one year. The vehicle was originally a Midland brake van, as the brakeman's 'ducket' will show.

D. Thorpe

(Above) Cashier's coach No. 3, with windows boarded against vandals, stands in Mode Wheel yard in June 1980. The coach, which was originally a horse box, was the last of a series which carried the cashier's and guard from Salford to Latchford. The 'coach' was sold for preservation and left the Manchester Ship Canal for 'Steamport', Southport on Friday 7th January 1983.

D. Thorpe

An ex-L&YR coach at Salford yard in 1965. The coach, which was a fireproof vehicle, was attached to the Resident Engineer's Department.

R. S. Greenwood

Ellesmere Port Engineer's breakdown van No. 6718. an ex-GWR 'Toad', the platform was enclosed for weather protection after being purchased by the Company. Note the large timber baulk (jacking supports) on the white painted step.

D. Thorpe

the Manchester Ship Canal system. The first to arrive was *Ranger*, a four-wheeled brake of uncertain vintage and origin. It was purchased from the Standard Railway Wagon Company Ltd. for the sum of £500 and arrived at Manchester on 22nd June 1951. After an internal rebuild and the fitting of a NE type brake system, the coach emerged from Mode Wheel as a kitchen/mess room vehicle for use by the platelayers. The exterior was painted the usual red oxide with the letters MSC Co., RED and the name *Ranger* painted in white on both sides of the coach.

In 1965 the second coach arrived at Manchester, and was destined to serve in a similar capacity to *Ranger*. The vehicle was an ex-BR eight-wheeled brake third, No. 27001. Once in the shops, the whole interior was ripped out and the resultant shell was divided into two unequal sections, the smaller section being converted into a kitchen where butane gas was used for cooking and heating. The outside was painted in all over green, of a rather indeterminate shade which was no better than the red oxide of the earlier rolling stock. The name given to the coach was *Rover*.

Breakdown Vans

For most of the years on the Manchester Ship Canal the means of transport used by the breakdown gangs was simply a goods van, usually a NE type, which was shared by men and equipment. An indication of just how risky it was riding in the van is given in Chapter 9 in the section dealing with breakdowns.

The last van of this type used for the purpose, was bought on 25th August 1944. It was given the MSC number 4283 and apart from a set of steps under the door, a few hooks and brackets were also fitted inside, on which to hang ropes etc. There was no other work done on the van to prepare it for its new role. After this van came MSC No. 5888 which was delivered on 7th April 1956, and on this occasion an attempt was made to convert it to a vehicle more suited for the purpose for which it was intended. It came to the Canal from Derby and cost £130. Amongst the alterations carried out on the van, was the fitting of a steel plate on the floor by the side doors, tool locker seats, a cast-iron stove, and lifting equipment to aid the hoisting in and out of the van of such items as the traversing jacks and the wagon ramps. Brackets were fitted to hold the equipment in place whilst the van was moving. Timber baulks and plates were stacked at the opposite end from where the men would be seated. For a while the old van was used to carry plates and baulks.

A much more suitable form of transport was adopted when, in August 1962, the Company bought 3 ex-GWR brake vans. This type, usually classed as 'toads', were by far the most adaptable to the requirements of the breakdown unit. The open veranda at one end was perfect for storing timber baulks and plates, whilst the closed in cabin could be laid out in a very tidy manner with lockers and cupboards, and with securing brackets for lamps and jacks. The usual cast-iron stove was fitted. The fact that it had windows at each end was another point in its favour, for to ride in the earlier windowless vans, with the side doors wide open, on a freezing winter day was not really pleasant. Both other brake vans went to Ellesmere Port where they were used in their original role. The situation was improved further with the arrival of a fourth ex-GWR brake van on 17th April 1970. This van, No. 6719, was obtained for the purpose of housing the Maschinenfabrik Deutschland Department equipment and was obtained from BR Margam at a cost of £350. Ellesmere Port also received a van of the same type, the MSC number being 6718. At the former open end of the van it had been fitted with all-round window frames, and this vehicle was put to use as a breakdown van.

Redundancy of Rolling Stock

With the closure of the docks railway system in 1972, the major part of the Company's rolling stock was made redundant. On 26th January 1973 a grand round up of wagons began, the whole operation being under the direction of Chief Superintendent Inspector John Drinkwater. By 7th November a total of 1,140 wagons of all types had been gathered for scrapping, most of them going to Taylor Bros. in Trafford Park. There were odd pockets of wagons remaining which were left for local use. A good number were left at Ellesmere Port, whilst at Salford, on a completely isolated section of track, a train of 'Conflats', (bolster wagons converted to container carriers) and bogie bolsters, were put to use as container trolleys. These trolleys were simply used to take containers to or from the ship's side, and to or from a storage area. They were moved about by Muir-Hill shunters.

Out of all the thousands of items of rolling stock which formerly worked on the Manchester Ship Canal system, very few remain and even less are in the hands of preservation societies.

Chapter Nine
Workshops and Breakdown Facilities

The Mechanical and Electrical Engineer's Department is responsible not only for the buying and maintenance of railway equipment, but for all manner of machinery and plant, which ranges from the smallest piece of electrical equipment to the overhaul of lock gates, tugs and dredgers. There are several workshops, some, like the Gate Shed, only dealing with one particular aspect of the work, whilst others deal with a multitude of jobs, usually at local level. Of these shops there are a few that are, or have been, concerned with the maintenance of locomotives, rolling stock and associated equipment.

When the Manchester Ship Canal took over the docks of the Shropshire Union Railway & Canal Co. at Ellesmere Port, it took over a railway system known locally as the 'Shroppie' and an engine shed, which was situated near the town's railway station. The shed formerly housed the dock branch engines of the LNWR. In the photograph, taken on 22nd January 1949, MSC 'Austerity' locomotive No. 86 stands outside the shed, raising steam. Over the years, the building had suffered many indignities such as a locomotive overshooting the stops and bringing down the water tank at the rear and, on another occasion, having one set of doors smashed when No. 85 burst its internal steam joint to the cylinders, resulting in a driverless runaway.
A. Appleton

Ellesmere Port originally maintained the engines at the locomotive shed near the station with a back-up service from the workshop on the docks. 'Out of the ordinary repairs' were, and still are, handled by fitters from Mode Wheel shops. After the shed was demolished in 1972, maintenance was carried out at the new workshop, which had been built two years earlier, a fitter and one electrician being sent from Mode Wheel to service the locomotives.

At Runcorn the Bridgewater Department have their own workshops which, up until 1952, handled all but the heaviest locomotive repairs. Locomotives working at Acton Grange were serviced by the fitters from Latchford Shops but, like Ellesmere Port and Runcorn, were attended to by the Mode Wheel staff on the larger jobs.

Latchford Shops, if not an ex-contractor's shop, was certainly an original Manchester Ship Canal building. Of timber construction with a felted roof, it was typical of those used by the contractors when the Canal was under construction. Another workshop of the same type was one at Partington Coaling Basin, and until 1938, quite heavy repairs to locomotives were handled there, including major boiler work such as new fireboxes etc. Partington Workshop was replaced by the present building in 1964 as it had become structurally unsafe.

The early locomotive repair shops are not well recorded and in fact only one such establishment is shown as existing in the six years up to 1900. This was the engine repair shop on Trafford Wharf, which was also used for general repair work. In Chapter 2 mention was made of the 'new' workshop at Mode Wheel, and it was stated that the 'old' workshop was pulled down and the ground cleared for traffic purposes. It seems highly probable that the 'old' workshop was, in fact, the one on Trafford Wharf. The contract for the building of the new works was placed in June 1899, and twelve months later the shops were in use.

With the opening of Mode Wheel workshops, in 1900, the engineering capability of the Mechanical Engineer's Department was increased by a tremendous amount. The up-to-date and heavy capacity machine tools meant that they could take on work which previously would have had to be sent to outside firms. As the years passed, the quality of the work coming out of Mode Wheel, and also the quality of the soundly trained artisans, won the respect of other Manchester Ship Canal departments. Not only were the services and expertise of the men called upon to deal with major locomotive repairs at other sections along the Canal, but it was also requested for general work in other departments. Mode Wheel shops was and still is the Company's main mechanical engineering works. The plans show its location and layout. Around the largest building, the fitting machine shop, is grouped the smaller buildings housing the boilermakers, tinsmiths, blacksmiths, joiners and wagon makers. The 'sock'-shaped building contained the main offices and the time lodge. They were constructed of timber and housed not only the Workshop Superintendent and his staff but also the Locomotive Superintendent. In the building over the other side of the works, which contained the pattern shop and the tinsmiths' shop, was an office for the Locomotive Foremen. Nearby was the locomotive shed, the gas plant and the sand drying furnaces.

In both the wagon shop and the fitting machine shop, lines were laid down each side. The lines in the fitting shop were equipped with pits for under locomotive work. About half-way down the shop, the lines converged to a single line which then continued most of the way down the centre of the machine shop. A set of lever-operated points was installed at the intersection. A single line was also laid into the paint shop, and one was laid between the locomotive shed and the smithy. The latter line served the chain shop, sand dryer and joiners' shop as well as the smithy. Running between the line and the single line section in the machine shop was a short length of line which connected to neither of them. It passed through the boiler shop and enabled boilers, which had been lifted from the locomotives in the fitting shop, to be moved into the boiler shop on a trolley made for that purpose. The boiler shop was

An inside view of a section of the present Ellesmere Port Workshop taken in June 1978. No. DH3001 was, at that time, in for servicing, this being carried out by Mode Wheel staff.

D. Thorpe

Latchford Shops, Warrington, with the ex-LNWR viaduct in the background. As can be seen, the building is of timber with a felted and tarred roof, and is typical of the contractor's building during the construction period.

D. Thorpe

An inside view of the new Partington Workshops, taken from the offices, showing the lights in the inspection pit. Another point of interest is the nameplate on the locomotive, showing that it was the first diesel-mechanical engine bought by the Company. No. 2001 *Ashdown*, later became simply No. D1.

M S C Co.

A view of the Fitting Shop taken from the overhead crane, possibly around 1931. In the shop at the time was locomotive No. 36 *Port Limon*, a Kitson, and Hudswell Clarke engines, Nos. 60, 57, 64 and 63. Note the boilers in the shop in process of repair. The boiler shop at that time was very small and work was often done outside that particular department.

Courtesy A. Appleton

A Smith & Rodley steam crane at Partington Workshops in June 1978. Formerly the yard crane at Mode Wheel Workshops, until it was displaced by a diesel crane in 1963, the four-wheeled steamer had led a varied and useful life. This crane is the only example of a rail-borne steam crane in use on the Canal.

D. Thorpe

A derailment at Trafford Park in June 1978, involving a train of billets for British Steel (ex-Taylor Bros) and a Rolls-Royce Sentinel locomotive, No. 3003, (out of the picture). The breakdown unit, headed by a diesel-hydraulic engine, is in two sections allowing road traffic a clear passage. The wagon is being lifted, using the older method of hydraulic jacks and timber baulks instead of the MFD equipment. In cases like this, it was quicker and less troublesome to ignore the MFD equipment.

D. Thorpe

extremely cramped and possessed no overhead crane, therefore to manoeuvre the boilers into the various positions whilst repairing them was a real feat of skill. Later, when larger boilers began to be repaired, the heavier work was carried out at the end of the fitting shop.

The smithy was equipped with two steam hammers, one being manually controlled, the other semi-automatic controlled. A great deal of the smithy work came from the wagon shop. At that time the wagon shop held four standard wagons;, but there were also repairs which were carried out in the yard outside the wagon shop. Behind the smithy was the chain repair shop and chains were brought in from all sections of the Canal. They ranged, in size, from ¼in. section chain to 2in. section chain, and occasionally even larger sections were handled.

A major change in the layout of the shops was seen in 1937. A new boiler shop and paint shop were built on the site of the old offices, a main stores with canteen and offices above was erected alongside the western wall of the fitting machine shop, and a new chain making, testing and annealing shop was built at the rear of, and protruding from, the smithy. An extension was built on to the wagon shop, giving space for a further four wagons. The time lodge was replaced by a building of brick construction, with a timber built cycle shed adjoining. Each of the new shops and also the stores had an internally laid line.

The gas engine driven compressor was replaced by an electrically driven, single cylinder horizontal unit, made by the Consolidated Pneumatic Company. It was housed in a separate building inside the boiler shop. Shortly afterwards the gas plant was made redundant and was eventually demolished. In the boiler shop a 10 ton overhead crane, manufactured by Vaughans Cranes Ltd., was installed. It was obviously a lot more advanced in design than the fitting shop crane, and was a great deal faster. The first foreman boilermaker in the new shop was a Partington man, Joe Morgan. With the larger working area then available, the way was opened up for larger jobs such as steel fabrications, and there was also more storage area for repaired and new boilers.

With the removal of the gas engine opposite the section then occupied by the tinsmiths, a department was set up for the rewinding of motor armatures. The men employed in that section were, in fact, involved in very much more than armatures. Most types of electrical controls and other jobs were, and still are, dealt with in the shop. All the coils for the motors are made in the department, the sizes of the motors ranging from fractional horsepower units, up to 40 horsepower units. With the introduction of diesel locomotives, the electricians, or it should be said, electrical fitters, were drawn into the work of maintaining the new engines. There was a considerable amount of work involved especially on the diesel electric locomotives.

Although a certain amount of work was carried out on the locomotives by the joiners, the vast majority of jobs came from outside departments, and works maintenance occupied a lot of their time. Work done on locomotives included cab floors, bunker shuttering, buffer blocks, also tool and oilcan boxes. Outside work included such diverse items as wheelbarrows and handcarts, including the making of wheels, a very skilled trade in itself, and on the other end of the size scale, portable buildings and mock-ups of bridges for new type tugs. A man had to be extremely versatile to work in any of the Company's workshops.

In the early days there was room in the fitting shop for six locomotives, three on each road. Engines which did not require the use of the overhead crane were dealt with in the running shed by the fitters, or boilermakers, if it was in their province, who were stationed there. Working conditions in the locomotive shed were quite bad and there were very few facilities to make the job bearable, just a bench along one wall near the bottom of the shed, only the crudest of equipment, but an abundance of choking smoke and running water. The same conditions lasted for over fifty years before any major changes occurred to make life 'down't shed' more acceptable. Conditions in the fitting shop were certainly a lot better, but the equipment was only slightly more up-to-date. Even so, in those early years, it was as good a shop as many of the workshops of the locomotive builders from which came the vast numbers of engines built for countries all over the globe, and much better than some workshops belonging to other private companies. One of the most used pieces of equipment in the main shop was the overhead crane. Built in 1900 by Craven Bros. of Manchester, it was of plate girder construction and had a capacity of 15 tons. The top speed of lift and travel was not very fast but it was sturdily built and lasted over 70 years before being replaced. Actually the last 34 years of its life was spent as a 12 ton unit because of the down rating due to 'blitz' damage to the workshop walls in 1940.

Boilermakers repairing a Priestman level-cut grab. Grabs were a regular part of the work in the Boiler Shop.

Courtesy J. Kay

For many years a crossing-smith was employed at Mode Wheel shops and although he operated from the smithy, the very size of the completed crossing meant that the final assembly had to be carried out in the larger area of the Machine Shop with the facility of the overhead crane. The diamond crossing shown was ready for handing over to the permanent way staff. An interesting detail to be seen is the early belt-driven wheel lathe to the left of the planer. Both machines were made by Craven Bros. Ltd., Manchester, circa 1900.

Courtesy A. Johnson

Matterson electric lifting units in action. These were purchased in 1960 to deal with the heavier and taller engines which had entered service. The existing overhead crane and track support masonry were judged suspect, due to war damage, but the cost of renewal, plus the cost of erecting a higher roof, at £64,000, was said to be too costly. At around £7,000, the electric jacks proved to be a cheap way of solving the problem.

Courtesy Mattersons Ltd.

The Fitting Machine Shop, around June 1951, viewed from the overhead crane. In the Fitting Shop at the time was (left) Hudswell Clarke long tank, No. 68, whilst behind is another locomotive of the same class, No. 64. To the left is 'Austerity' locomotive No. 85 and to the rear is 'Jazzer', No. 16 *Galveston*. In the foreground of the Machine Shop is the 11 ton 'wedge', one of the pair used in the sealing of the ends of Barton Aqueduct.

R. Johnson

Locomotive engineering practice varied very little from one builder to another, and in many cases their methods of manufacture had advanced only slightly over the years. Of course, in some cases their product also remained much the same; uncomplicated, unsuperheated locomotives, which is all that most industrial users required. The last steam engines of the Manchester Ship Canal Company differed merely in size, appearance and certain minor details from their earliest engines, with Stephenson link motion and dee-valves, and straight-slotted ports, slide bar and crosshead arrangements, following standard practice and, in fact, like most other items on the engines, were little changed since the early days of steam. It is not surprising, therefore, that certain methods used in repairing the locomotives had also remained unchanged at Mode Wheel since the earliest days of its existence. Jobs like the relining of axleboxes using steel plates and copper rivets, facing horn-cheeks and reseating cylinder valves. The latter job was a particularly soul destroying task, usually taking up to three weeks to complete by one man. Whether the locomotive had an inside or an outside cylinder arrangement, it was still a long wearying chore.

Another process which remained unchanged for many years was the retyreing of locomotive wheels which, after being machined internally to a shrink fit, would be expanded by heating in a wood fire. Once the tyre was placed on the wheel it was contracted by means of a circular water spray ring. Tyreing usually was started early in the morning, the bonfire being started long before normal working hours. In this way the correct expansion could be reached by the time the chargehand was at work, and only on extremely wet or windy days was the job halted. An alternative method of heating the tyres was tried using a works-made gas ring, but this did not prove successful and the idea was abandoned. However, in 1962 a proprietary make of gas-air equipment was tried and found to be slow but adequate for the job, with the added bonus that it could be used in the shop. Apart from the gas-air

A front view of a Kitson, with cylinder and valve chest covers removed. To reface the steam chest valve faces was a three week job for one man using long-handled files, scrapers and pressure levers. It will be seen that with a 7in. width aperture to work in, the task called for a great deal of patience and expertise.

J. Steggles

mixing machine itself, the system used several cast-iron burners which were grouped around the tyre, the larger tyres taking more burners. Cooling the tyre is supposed to be accomplished using air only coming from the same blocks, but to speed up the process water is sometimes used. Newly-tyred wheels are then placed in the wheel lathe and the tyres machined to the correct size and profile. In the photograph showing the diamond crossing, the early wheel lathe can be seen in the background. It was a Craven Bros. machine, built in 1900, and driven from its own motor through large width flat belts.

Locomotive fitting is a heavy, and very often, dirty job, but it is something, with all its drawbacks, that the staff take in their stride. Today it is the clinging smell of diesel fuel, but in the past it was blastpipes and main steam pipes in sooty smokeboxes, the adjusting of big and little ends with shoulders rubbing against oily motion and with feet paddling in wet pits, but it all added up to a job which was done without too much face-pulling. However, one job that was not very popular was the work which had to be tackled inside the water tanks. If the tanks belonged to a locomotive newly-arrived from Ellesmere Port, then the job was even less welcome. It was not claustrophobia which put people off (although it has happened), it was the pond life which lurked in the remaining water at the bottom of the tanks. The wriggling fishes were startling enough, but the horrible looking beetles and other slow moving creepy crawlies which appeared in the flickering light of the candle, put even hard-bitten types to flight. Apart from the 'creatures of the deep', conditions in the tanks were generally unpleasant, and the doubtful privilege of working in them was usually given to the apprentices.

In a more serious vein, in the earlier days, however unpleasant or rough a job turned out to be, there was no dodging it, for most of the men were on piece-work and if a locomotive was delayed a man's wages would be down. Piece-work rates were very closely worked out and the men had to be fast, but thorough, workers to earn even a decent wage.

To take the light overhaul of a 'Jazzer' as an example, in 1933 the total cost was worked out at a mere £14 7s 1d! The largest single payment was for the stripping of the locomotive which would earn the fitter and his mate £1 14s 0d. Woe betide a man who had been careless with his work, the rate for adjustments after trial being a princely 2s 8d. There were, however, allowances for extra work which had to be done, 7s 4d to bed a brass into an axlebox, and if an engine was stripped, including the boiler removed, the price was £3 0s 0d. Boilermakers could earn £39 7s 7d for one job, but to get that sum they would have to renew a firebox, including a complete retubing of the boiler, and most of the work was done using hand tools. Blacksmiths would have to forge six Hudswell or Hunslet crankpins in order to earn 7s 0d but a single crankpin for one of the early engines such as *Irk*, *Bolton*, or *Romilly*, would be worth 1s 6d . Admittedly, the pound (£) was worth more at that time, but it was a hard way of making a living.

The second wheel lathe at Mode Wheel shops which replaced the earlier Craven Bros. machine in 1950. Built by Tangey Ltd., a firm better known for pumps and hydraulic equipment than machine tools, the machine was of simple, but robust, construction and is an extremely useful general purpose lathe.

D. Thorpe

A planing machine in the Machine Shop at Mode Wheel, in 1978. Apart from the many varied general engineering jobs which are brought in from all sections of the Canal, the machine, built by Craven Bros,. Ltd in 1900, was used to taper the rail sections used in crossings and points.

D. Thorpe

However, to be unable to make much money out of the piece-work rates was bad, but the spectre of the sack for being slow on the job was far more serious. The Company were not ones to carry 'passengers' and instant dismissal was always a possibility. Apprentices were obviously given more latitude, but they were faced with the depressing certainty of dismissal when they finished their time and reached the age of 21. After finishing his apprenticeship a tradesman would have to take a job elsewhere as a 'journeyman' or improver, a position he had to hold for at least one year before he was considered to be fully proved. If he was lucky, he would then be able to start back again as a fully-fledged tradesman and this practice was common to many firms in years gone by. Actually, at Mode Wheel it was often known for a man to be sacked on his 21st Birthday then be re-employed the following day as a journeyman, thus following the customary practice of keeping a good man for their own benefit.

With the coming of World War II, the piece-work system and the practice of sacking at 21 was discontinued. With regard to the piece-work, it is interesting to note that the length of time taken for overhauls and rebuilds took no longer, on average, in the 1950s, than it had done years earlier. A rebuild would take anything between 8 to 12 weeks, depending on the extent of the renewals. Light overhauls could be between 2 and 4 weeks, whilst heavy overhauls took about 6 to 8 weeks.

Locomotives from outlying stations requiring overhaul were, at one time, sent to Mode Wheel by rail. For the Partington engines this was a simple matter, but for those at Acton Grange, Runcorn and Ellesmere Port it was a more involved operation. The complication was that the motion and coupling rods had to be stripped and fastened securely in the cab, for the locomotive was taken, as a wagon, on a slow goods train over the metals of the LNWR. Due to the damage done to the axle bearings on several occasions, the Manchester Ship Canal decided, in the early 1950s, to send the locomotives by road.

Naturally, the most satisfactory part of the overhaul, or rebuild, of a locomotive was a successful trial. Even to those who had been responsible for many major locomotive repairs in their time, there was still a feeling of pride when a trial engine returned to shed with a verdict of 'no further work necessary'.

The day before the locomotive was drawn out of the shop would be occupied in final small detail work such as cylinder drain cock settings, making sure that cab pipework was connected properly and of neat appearance, and fitting whistle chains and brake gear safety chains etc. On the day of the trial, the engine would be fired early in the morning so that steam would be up in time for the chargehand fitter to set the safety valves early in the day. An apprentice fitter would be told to report to the engine with the chargehand's tools; how important the lad would feel climbing into the cab ready for the excitement to come. The engine would be standing outside the shed looking quite smart and clean, with the boiler

Kitson locomotive No. 72 is seen being taken through the main street at Ellesmere Port, on a Pickfords low loader, on Thursday 7th June 1962. The engine was bound for Mode Wheel Workshops where it was to be overhauled. In past years, the practice was to send the engines by rail as a wagon, with motion and coupling rods removed, but for various reasons, one being the damage to the axle bearings caused by excessive speeds, the Manchester Ship Canal decided to send them by road.

D. Brown

Kitson locomotive No. 72 is seen being unloaded from a Pickfords low loader on Trafford Wharf. Afterwards the engine was towed to Mode Wheel Workshops where it was due for overhaul.
Courtesy Mrs Loddington

simmering gently as the pressure built up. A glorious smell of hot cylinder oil and new paint would assail the nostrils. The driver would be checking the oil flow on each trimming, and checking the sight-feed lubricator whilst the fireman would be keeping himself busy with the job of getting steam up. All eyes were on the pressure gauge, the front windows pinned open, the pressure reading being given to the chargehand kneeling alongside the safety valves.

When all was ready, the locomotive crew, chargehand, apprentice, and sometimes the fitter who had been in charge of the engine's overhaul, would climb aboard and the engine would quietly move out of the yard and on to the 'route'. As it ambled along the line and through Weaste Junction, everyone would be listening for strange noises and sniffing for overheated bearings. At No. 29 Stores, the bonded sheds, the line up to Irwell Park Wharf runs dead straight, and for about 700 yards is suitable for testing a locomotive at speed. At a point just before the straight section begins, the locomotive would be stopped whilst the chargehand fitter checked all bearings, as well as the joints on the main steam pipe, the cylinder covers and the steam chest covers. After satisfying themselves that everything was in good order, everybody would climb aboard for a fast run.

Now there were some drivers who did not agree with the practice of taking the engine up to the speed at which it was tested, arguing that an engine would never be called upon to do that sort of thing in its working life. However, the majority of drivers did exactly what was expected of them and opened the regulator wide. The sight of an industrial engine rocketting its way down the straight might not be a main line enthusiast's idea of a spectacle, but to those who were used to seeing them rumbling slowly along with a train or gently puffing along 'light engine', it was definitely a sight worth seeing, especially if the engine was a Kitson or one of the other larger engines.

The services of the men at Mode Wheel were often requested by the superintendents from other sections, the fitters and boilermakers having to go to Runcorn or Ellesmere Port for locomotive work and also to other sections such as grain elevators at Salford and other similar jobs. Usually when

Locomotive No. 30 *Kurrachee*, Hudswell Clarke No. 663, was built in 1903. The engine, which had just been rebuilt at Mode Wheel shops, is seen standing outside the engine shed, on 8th September 1948, resplendent in the new livery of light grey, edged in black and lined with white. Seated on the engine is apprentice fitter Alan Locke, whilst the fitter responsible for the rebuild, Harry Elliot, stands on the far side of the buffers. Leaning on the buffer is Fitting Shop labourer Arthur Hambridge.
Courtesy Alan Locke

A view looking westwards towards Irwell Park Wharf and Barton Bridge. The line directly facing the camera was the 'long straight' used by Mode Wheel Fitting shop staff for the fast run during locomotive trials. The Canal is on the left of the photograph.

D. Thorpe

a fitter had to go to a locomotive job, it was one which was considered outside the scope of the local people. That is not to say that the local men were not capable of sorting the trouble out, it was because of the time scale in working out the method of doing the repair, as against the experienced approach of Mode Wheel men who obviously were familiar with every aspect of locomotive repairs.

On one occasion the fitters were sent to Ellesmere Port Shed to look at an 'Austerity' which had developed an internal leak somewhere in the main steampipe. The trouble manifested itself in a very dramatic way and it was a miracle that no one was hurt in the process. The locomotive was in the shed whilst steam was being raised and it was bitterly cold; the main doors were closed. The shed staff were talking to a BR driver who had just clocked on, an arrangement between British Railways and the Manchester Ship Canal, when a gentle sigh was heard. As the startled men looked round they were in time to see the engine slowly move off in the direction of the main doors. Before anyone had time to act, the locomotive had smashed its way to freedom and was loping along the track towards the BR exchange line. The runaway was boarded before it got too far, and was bought back to shed. When the people at Mode Wheel were informed, they asked the Ellesmere Port driver to carry out the usual tests so that the trouble could be pinpointed, and the result was a cone joint failure. Two fitters were immediately sent to the 'Port' with the necessary equipment. When they arrived at the shed, the locomotive was ready with the boiler drained and the dome cover removed. As was stated previously, the weather was bitterly cold and this led to a tremendous contrast in working conditions for the two men. The fitter inside the hot boiler with his legs straddling the tubes, was streaming with perspiration whilst the man working the tommy bar, unfastening the regulator header bolts, was continually being forced to leave his job to totter into the locomotive shed where he could 'unfreeze' himself. The job took less than two hours to do but nearly lost the Company two fitters, one by heatstroke, the other by hypothermia!

Locomotives belonging to outside firms were also worked on by the men at Mode Wheel. In some instances it was a derailment and in others it was a minor repair, but the overhaul of other firms' engines was only done on rare occasions. The Peckett which belonged to Lords Tar Works at Weaste, was given a complete overhaul during 1941 and was given a light overhaul and boiler repairs in 1944.

Breakdown Facilities

Derailments were not uncommon especially in the dock system where debris sometimes dropped between the blades of the points and filled those lines which were laid flush along roads and quaysides. In the latter case the drivers were adept in driving themselves back on to the line, and on some occasions even where the line was laid in the usual way and not laid flush. Trafford Park with its complex system was another happy hunting ground for the breakdown gang.

Some derailments, however, demanded very much more than a few wooden wedges and pieces of metal plate. An engine would often end up several feet away from the line and, in some cases, tear up the track in getting there. In these instances it was a job for the breakdown gangs. The gangs were based in sections very similar to the platelayer's gangs,

The results of a bad smash to the blue and red fireless engine of Proctor & Gamble Ltd. at 'B Junction', Trafford Park, after the engine, a driverless runaway, had careered along Trafford Park Road for almost half a mile, crossing two road junctions before hurtling into the sidings to collide with the MSC engine No. D12. The Manchester Ship Canal engine was knocked back about 20ft. and sustained a broken buffer, but luckily nobody was in the cab and there were no casualties. The accident occurred on 24th June 1964.

J. Steggles

LMS insulated van No. 67807 at cross purposes with the track. It is a source of amazement how wagons and locomotives involved in derailments sometimes end up in positions like this. The picture was taken sometime during the 1930s.

Courtesy J. Steggles

with whom they occasionally worked. At Ellesmere Port and Salford they were equipped with vans, and at a later date with an additional unit, a brake van. The snag with the earlier vans was that the crew had to ride inside amongst the equipment. On some occasions, when the locomotive hauling the van came to a sudden stop, the men would be thrown on to the gear, or even worse, the gear has been known to fall on to the men. The trouble mainly lay with the large quantity of timber baulks which had to be carried, as any poor stacking of the baulks was sure to bring the whole lot tumbling down.

There were no derailment cranes on the Canal, simply jacks, ramps and baulks. The ramps were used in some of the wagon derailments, but there were times when even the wagons had to be carefully manoeuvred using the lifting and traversing jacks. The danger with some wagons was the possibility of smashing them whilst lifting, for the wagon involved may have been one of those used extensively for carrying sulphur, a commodity which did very little good to timber built wagons.

To put a locomotive back on the track after a derailment involving all wheels off, was a fairly straightforward task, that is if the track was still in a good enough state and if the engine was not leaning at a crazy angle or, again, if the ground was reasonably firm on which to work. The reasons why it was not always straightforward could fill a book, but even so, the breakdown gang always had the answers. On some occasions the line was so badly damaged, the locomotive had to be moved over to the opposite track, if there was more than one set of metals. On other occasions the engine had to be lifted so that the platelayers could build a temporary track, in order to move the locomotive clear of the affected area. The platelayers proved time and again to be very co-operative on such breakdowns.

There have been several unfortunate incidents on Bridgewater Bank, leading to Bridgewater Junction, one ending with the locomotive on its side and a shunter receiving such serious injuries that he was unable to handle the work of the shunter, and was consequently given the job of crossing keeper at Salford. The locomotive involved was a 'Jazzer', believed to have been No. 21 or 22. On this occasion the engine was put back on the track by the LMS breakdown crane which had been brought in from Gorton. On another occasion No. 4002 was drawing a train out of Junction 'B' when one of the wagons started to break up. The load, which was composed of metal ingots, dropped through on to the track. As the wagons following smashed into the ingots, more and more of them were damaged, and at the same time the track was being torn up. Such was the power of the diesel-electric locomotive that the driver, who could not see the major part of his train due to the bend, did not feel any hesitation on the part of the engine and only stopped when a frantic shunter managed to catch up with him. Although there was debris and parts of wagons scattered all over the place, the wagon inspector managed to sort out the pieces into their correct piles, each pile representing a wagon, all that is, except one. The wagon was never entirely found, just odd parts. The wagon which disappeared could possibly be No. 3121A, which mysteriously appeared on the docks in October 1953 and which had gone back to the nether regions from whence it came!

A breakdown gang was made up of men who had normal day jobs in or around the workshops, and who were on call during the night or at weekends. The job of informing them of a derailment during their off duty time was, for most of the men involved, rather difficult, as very few of them had telephones at home and, therefore, the weary task fell on one of the unfortunate shed lads who would have to cycle from house to house rousing the men. Many a lad had come back

Derailment on Trafford Wharf with Hudswell Clarke locomotive No. 62, its nose into the ground. The cause of the accident, a pile of gravel heaped under the first tank, had been dropped from a hopper wagon in a preceding train when the bottom doors gave way.

J. Steggles

A breakdown unit at Mode Wheel yard in 1979, comprised of two ex-GWR brake vans (Toads). The van which is marked MFD carries the rerailing equipment manufactured by the German firm Maschinenfabrik Deutschland. The MFD equipment comprised a light alloy beam, lifting jacks, traversing trolley and hydraulic traversing units. Power to the hydraulics is supplied by a petrol engine-driven pump, control being maintained through a portable console.

D. Thorpe

MFD equipment in use. To the left of the picture a lifting jack can be seen, whilst on the beam itself is the hydraulic traversing unit which is coupled to the traversing trolley. The locomotive buffer beam sits on a block of wood on top of the trolley. This particular derailment occurred during a summer night in 1978 at the crossing on Park Rd. near the Kellogg Plant, and involved the container base train.

D. Thorpe

Locomotive No. 4001 in shops for its first complete overhaul in October 1963. The 7½ ton National Gas & Oil engine Brush Traction power unit is being settled on to timber bearers ready for a stripdown. Chargehand Alan Locke inspects the points of contact, watched by Fitter Alan Johnson.

R. Johnson

to the shed, red-faced after having roused the wrong householder.

As can be imagined, the task of putting a locomotive or wagon back on to the track was often a hard and messy process. If the weather had turned foul, the icy rain or snow driven along by a gale force wind, and the ground churned up into a quagmire, it could become a nightmare, especially if it occurred during the night with the derailed vehicle at a bad angle. It was a good man who could take that sort of treatment and still remain on the breakdown gang. To add to the misery, in the early breakdown vans there was no form of heat and no facilities to make a hot drink.

A useful addition to the breakdown equipment was made in 1970 when the Department bought a new type of rerailing system. Maschinenfabrik Deutschland was the tongue twisting name of the makers of the equipment but it is normally referred to as MFD. Basically the MFD equipment consisted of an alloy beam which spanned the track and supported a system of hydraulically-operated jacks and a hydraulically-operated rolling table. Power is supplied from a petrol engine driven oil pump which is connected to a portable control console. Although the system still entails the use of jacks and traverser, together with the fact that it all has to be manhandled in much the same way as the earlier equipment, it has proved to be extremely useful.

Rolls-Royce Sentinel 300hp 0-6-0 diesel-hydraulic locomotive No. 3003, stripped down for overhaul during 1977. The wheels, which are made complete with flange, had been machined ready for the fitting of tyres.

D. Thorpe

The White House

After the war years, at Mode Wheel workshops, there was a considerable amount of unrest amongst the boilermakers and fitters, concerning the working conditions in the locomotive shed. These conditions have already been remarked upon and there was no question about it, they were disgraceful. If a man was engaged in work at the bench he was subjected to the damp, the cold and the smokey atmosphere. In his working day there was no place he could go to escape these conditions if he wished to dry himself or regain some warmth and to be found in the cab of a locomotive in steam was frowned upon. Although some of the older members of staff put up with things as they were, some of the younger men who were sent 'down the shed' as extra help, did not take to it kindly with the result that, in 1953, a new workshop was built for the shed staff at the bottom of the shed, which occupied an area formerly taken up by part of the 'sandhole' road. When finished, the new workshop, with its open fireplace at one end and its whitewashed walls, was promptly christened 'The White House'. Whether or not the United States Senate would have been flattered if they had known of this is doubtful, but there was no doubt as to the feelings of those who were to occupy the building; they were extremely pleased with it. True there were times when they were smoked out even after a new roof with centre vents had been fitted to the shed, but the directive to work down the shed was not considered to be such a savage sentence from then on, indeed there were even volunteers for the job. The staff consisted of a regular boilermaker, two fitters, three mates, and a boiler washer-out. It is interesting to reflect that the new accommodation, with a cheery glowing fire in the grate, received more frequent visits from the Superintendents than in days previously.

With the advent of diesel locomotives the departmental heads of the workshops had to change their policies with regard to tools. Previously the range of hand tools supplied by the Company was very limited and were issued extremely grudgingly. Consequently the workshops dealing with the new locomotives were allowed to invest in more suitable tools and equipment. At Mode Wheel a servicing bay was set up in the paint shop, a pit being installed as well as an exhaust extension pipe which fastened on to the chimney of the diesel-electric locomotives. This allowed the engines to be run up whilst the locomotive was still in the shop. It was a good idea but, unfortunately, it did not fit any of the other classes of diesel. When these engines were run up, it was usually done outside in the yard which was sometimes inconvenient but, nevertheless, better than suffering the nauseating fumes inside the shop. An exhaust duct was also installed in the fitting shop, but it has never been used for the same reasons which reduced the usefulness of the paint shop exhaust.

Services were based on the diesel engine manufacturer's recommendations, and for this reason the period between services differed between different classes of locomotive. For the diesel-electric locomotives with their National Gas & Oil Engine Company units, an 'A' class service occurred every fortnight, whilst for the diesel-mechanicals, with their Gardner engines, also the diesel-hydraulics with their Rolls-Royce engines, an 'A' class service occurred every three weeks. Class 'B', 'C' and 'D' services varied accordingly and the service requirements were set at a comprehensive level. If dieselization presented no problems to the fitting shop staff, it obviously meant the loss of work for the boiler shop. Apart from the phasing out of boilers, there was also a considerable amount of work lost from water tanks, bunkers and cab work. However, for some years previous to the introduction of the diesel locomotives, the boiler shop had been taking on more fabrication and plate work from other departments as well as repair work on grabs of various types. Each of the remaining shops involved in some part of the repair of steam locomotives, particularly the machine shop, also lost work, but obviously as none of their particular processes were as specific as that of the boiler shop, the introduction of other work was easier.

Chargehand Mike Tantrum checks over a Rolls-Royce 300hp diesel engine.
D. Thorpe

Apart from boiler work, Boiler Shop staff were involved in a diversity of jobs including grabs of different types, dredger buckets, general platework and fabrications. Pictured is a newly-completed catwalk for a lock gate.
Courtesy J. Kay

A busy scene in the Wagon Shop at Mode Wheel with a steel-framed box wagon in a stripped down state. The work was always very laborious with no mechanical handling equipment available, and to manhandle the various frame members often involved several men, including those from other teams.
Courtesy K. Hoare and G. Bullough

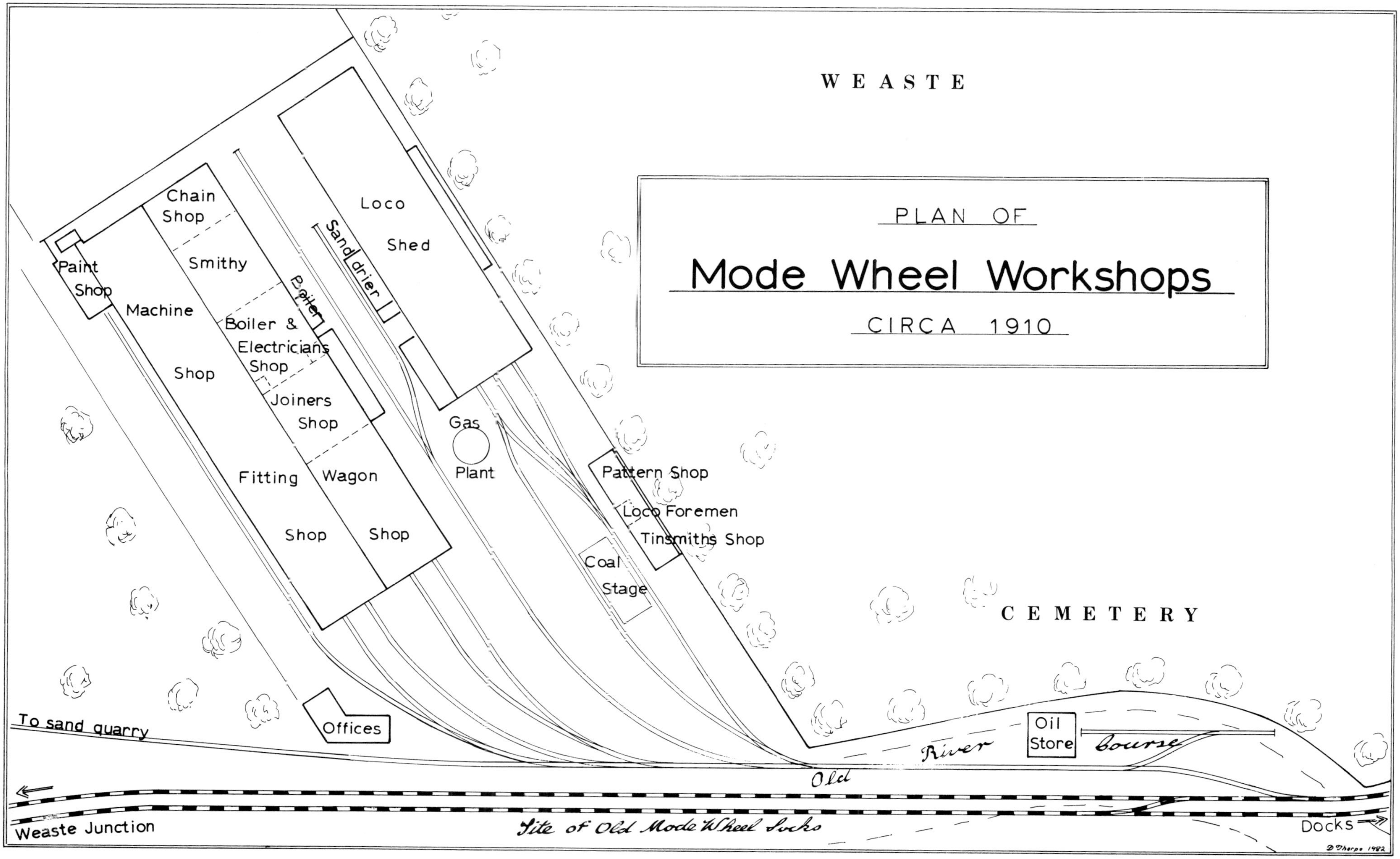

PLAN OF
Mode Wheel Workshops
CIRCA 1910
WEASTE
CEMETERY
Loco Shed
Sand drier
Chain Shop
Smithy
Boiler
Boiler & Electricians Shop
Joiners Shop
Paint Shop
Machine Shop
Fitting Shop
Wagon Shop
Gas Plant
Pattern Shop
Loco Foremen
Tinsmiths Shop
Coal Stage
Offices
Oil Store
To sand quarry
Old River Course
Weaste Junction
Site of Old Mode Wheel Locks
Docks
D Thorpe 1982

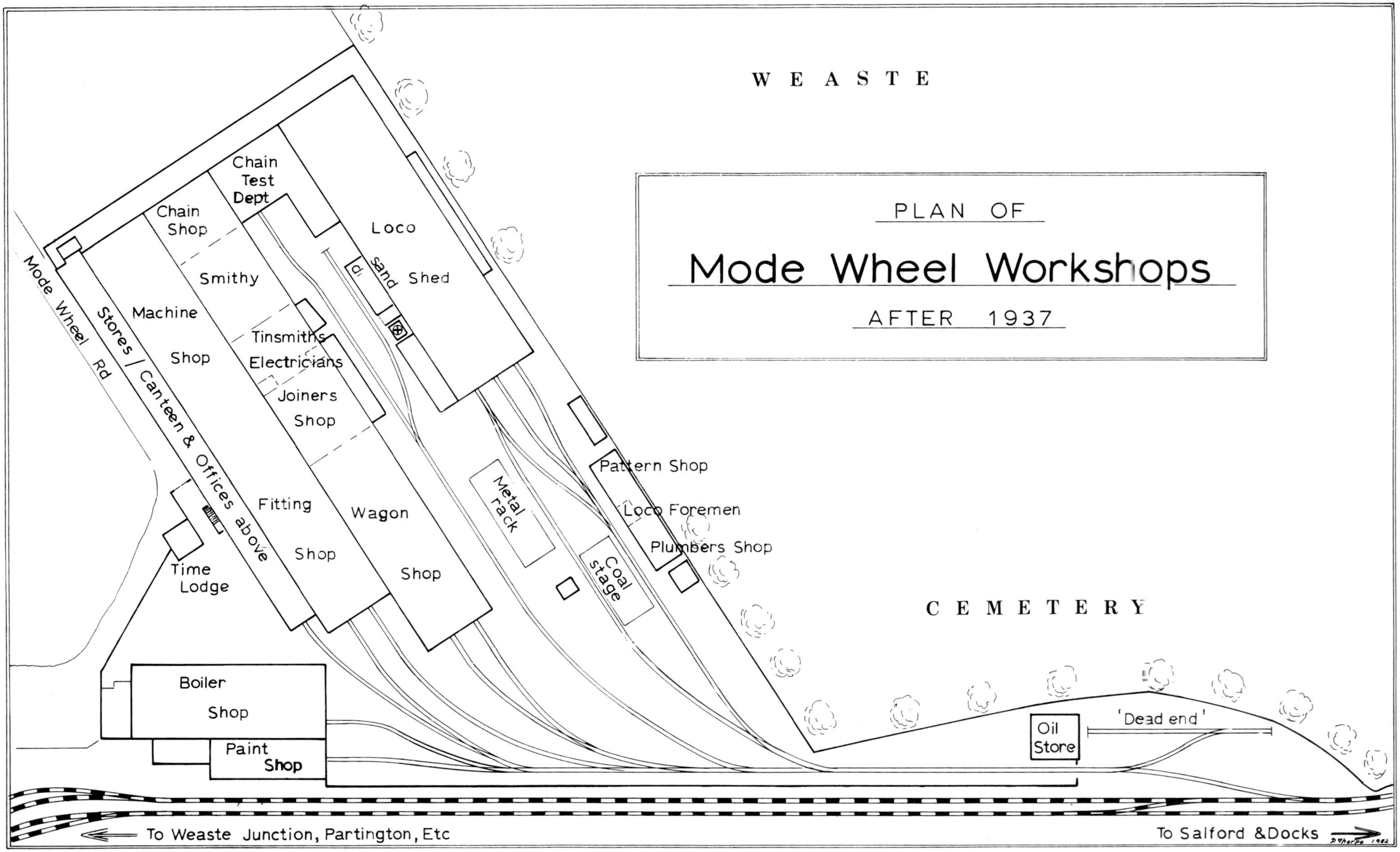
PLAN OF
Mode Wheel Workshops
AFTER 1937
WEASTE
CEMETERY
Chain Test Dept
Chain Shop
Smithy
Loco Shed
Sand
Machine Shop
Tinsmiths
Electricians
Joiners Shop
Fitting Shop
Wagon Shop
Stores / Canteen & Offices above
Mode Wheel Rd
Time Lodge
Metal rack
Pattern Shop
Loco Foremen
Plumbers Shop
Coal stage
Boiler Shop
Paint Shop
Oil Store
'Dead end'
To Weaste Junction, Partington, Etc
To Salford & Docks

An inside view of the new Barton Dock Workshop with locomotive No. D6 seen having attention to its axle brasses. This picture was taken in May 1982, soon after the workshop had come into use. No. D2 is just visible standing outside the building.

D. Thorpe

The new hopper for dispensing sand for the locomotive sandboxes at Mode Wheel yard. This replaced the sand drying furnaces which had been in use for over 70 years, since 1900.

D. Thorpe

The wagon shop, which was not affected at all in consequence of the change from steam to diesel, later received a real body blow with the closure of the docks railways. With practically the whole of the rolling stock being scrapped, or sold, there seemed no reason for its continued existence. However, keeping abreast of the changing times, the Company started the manufacture of certain items in glass reinforced plastic (GRP), and used the wagon shop for this purpose. Only a small percentage of the original work-force could be employed on this work, the rest were, therefore, employed in other departments, some of them going into the boiler shop and another man moved into the paint shop.

For the fitting shop, the docks railway closure did not have such a devastating impact as it did on the wagon shop, for over the years the increasing use of mechanical handling equipment on the docks, such as stackers and van loaders, had resulted in more repair work being brought to Mode Wheel. Other shops including the machine shop, the tinsmiths and the boiler shop also benefited from this situation. To ease the movement of this equipment between the depots on the docks and Mode Wheel, a road was built circa 1970 which came into the works yard at the southern boundary, and entailed the formation of a level crossing over the running lines.

Towards the end of 1983 much of the track was raised in Mode Wheel yard including the three shed roads. The vehicle maintenance department, 'Garage', as it is referred to, moved from Salford to Mode Wheel workshops, the boiler shop being made redundant at the same time.

During 1981 the Company decided to close the line between Mode Wheel workshops and Trafford Wharf with the result that locomotives working at Barton Dock Estate and Trafford Dock were unable to be serviced at Mode Wheel. Services were afterwards carried out on these locomotives at the new workshop, erected in the early part of 1982 on a site by the Kellogg plant at Barton Dock Estate.

Chapter Ten Preserved Items

Hudswell Clarke contractor's engine *Lord Mayor* (Works No. 402), built in 1893, has sometimes been credited with working on the construction of the Manchester Ship Canal, as it was delivered to Salford Docks in 1893. However, it would appear that this engine could have been destined for the site of the Manchester Dry Docks near Mode Wheel Locks. The engine was photographed in the exhibition hall of the Keighley & Worth Valley Railway, at Oxenhope.

D. Thorpe

It is interesting to look back on moments of history, and interest is sharpened when items connected with those moments can be seen, or touched, long after the moment has passed. However, where industrial organizations are concerned, the present and future are the important areas of time and very few firms take the trouble to retain their physical links with the past. The Manchester Ship Canal is, unfortunately, one which has not done so to any great extent, and consequently the Company has not preserved any major items of railway interest.

Before the fires of indignation burst forth from the breasts of the railway enthusiasts, sending them to march, in their thousands, to the doors of Ship Canal House, let it be remembered that to the Company the railway was but one tool amongst many, and to expect them to single out the railway as one section with items worth preserving is to be single-minded to say the least. There are so many sections, each with its own points of interest and each with its devotees.

To quote but a few, the Manchester Ship Canal has a fleet of superb diesel-electric tugs and, through the years, has had large numbers of steam tugs which ranged from the diminutive to the majestic. There have been, and in some cases still are, launches, work boats, floating cranes, dredgers, rail cranes, mobile cranes and road vehicles, and many more items of less obvious appeal.

To satisfy all bodies of opinion, the Company would have

Ex-Manchester Ship Canal locomotive No. 14 *St. John*, after being bought from the Manchester Ship Canal Co. by ICI Ltd. for its Blackley (Manchester) plant, was given a new livery and had the cab windows enlarged and was renamed *The Lady Armaghdale*. The engine worked at the Dyestuffs Division Works from May 1963 until December 1968, when it was sold for preservation to the Warwickshire Industrial Preservation Group.
Courtesy Mr Davies, ex-ICI, Manchester

The Lady Armaghdale, ex-MSC/ex-ICI locomotive No. 14 *St. John* at Bridgnorth on the Severn Valley Railway. The new livery of red, edged in black with yellow lining, suits the engine very well, the continuous line of the tank panels and bunker adding to the character of this grand little engine.
R. A. Panting

to set aside enormous amounts of money and space in order to house and maintain just one example out of each class or section. Just imagine preserving the 250 ton, or even the 60 ton, floating crane in the future. What a shed would be needed to house one of these!

However, in more serious vein, the Company has a great deal of money tied up in plant and equipment, and in order to maintain an up to date standard as economically as possible, the cost of the new has to be defrayed, in part, by the selling of the old, even if the amount received is only at scrap value.

There are, of course, people in the Company who would like to see more relics preserved and, in a modest way, it is done by the Public Relations Office, but funds and space do not permit them to form any kind of museum or gallery in which to exhibit large items. It is, therefore, left to private individuals and preservation societies to buy items of interest from the Company in order that these items are saved from the cutter's torch.

The first major piece of equipment to be saved was, in fact, bought, not for preservation, but to carry on doing a useful job at one of the I. C. I. plants. It was a locomotive, No. 14 *St. John*, and it was taken from Mode Wheel shops on 1st May 1963 to start its new lease of life at Blackley, Manchester. Once there it was overhauled and given a new livery. The main colour was cherry red, and it was edged in black and lined out in yellow. Regrettable, from the purists viewpoint, was the removal of the round ports on the front and rear cab panels and the substitution of 'D'-shaped windows. Actually they were more practical, it must be admitted, but all the same it is a pity that it had to be done.

Another alteration, which, although not unusual in itself, was curious in the way in which it was done, was to change the locomotive's name to *Lady Armaghdale*. The odd part of the change was that the original nameplate *St. John* was left in position on the inside of the cab rear panel; a case of split personality! The number plates were removed from the sides of the tanks and bolted on to the sides of the bunker.

After working at I. C. I. Blackley for just over 5 years, the locomotive was withdrawn from duty in December 1968. It was then purchased by the Warwickshire Industrial Preservation Group, and was transported to Bridgnorth in July 1969, where it was to run on the lines of the Severn Valley Railway. There were no further changes to either the locomotive or the livery.

The first engines which went specifically for preservation were No. 31, a Hudswell Clarke short tank, and No. 67, a Hudswell Clarke long tank. Both were purchased by Mr R. S. Greenwood and taken to the Keighley & Worth Valley Railway in June 1967. Both engines have been put to further use by their new owners, sometimes appearing at the head of the occasional winter train. They have also joined the ranks of locomotives which have appeared in films, No. 31 *Hamburg* suitably fitted with handrails, large headlamp, spark arrester chimney, and a bell, as a Russian tender engine in a television film. In fact No. 31 has appeared in several television plays being a very adaptable shape. Locomotive No. 67 had its brief moment of glory when it appeared in the opening scene of the delightful film, 'The Railway Children'.

Following the sale of Nos. 31 and 67, no further engines were sold until February 1973 when No. 32 *Gothenburg* was purchased by the East Lancashire Preservation Society, the engine subsequently being moved to the Bury Transport Museum for renovation. In June the same year, No. 70 was also bought by the East Lancashire Preservation Group and taken to the same home.

The first of the two locomotives to be renovated was No. 32 and although it was not painted in the original style, it was painted in a green very much like the original shade, and it was fully lined out with black edged with yellow. Another nice touch was the replacing of the nameplates on the side of the tanks. As there was only one plate to be found, that which was fastened to the inside cab rear plate, a second one had to be cast. The renaming ceremony took place on Sunday, 31st October 1976, and the honour of unveiling the plates fell to Mr S. A. Olsen the Managing Director of Ailsa Trucks (UK), the concessionaires for Volvo Trucks Ltd. At a later date a start was made on the overhaul of No. 70, much of the work, in fact, being done by an ex-Manchester Ship Canal fireman, Dave Smith. In typical enthusiast fashion, Dave taught himself new skills such as retubing boilers, and jobs of a similar nature, in order to see the locomotive in use once again. A deadline was set for completion in the early part of 1980 and, as a gesture to the Manchester Ship Canal Railway, the previously unnamed locomotive was to be given the name *Mode Wheel*.

Another engine which is of particular interest is the previously mentioned contractor's machine *Lord Mayor*, Hudswell Clarke number 402. The locomotive, which is preserved at the Keighley & Worth Valley Railway is an excellent example of the small standard gauge type which was to be seen on the Canal construction site. As it has been stated in Chapter Seven, *Lord Mayor* appeared on the Salford Docks site soon after leaving the makers on 7th July 1893.

Whilst on the subject of locomotives which did not actually work for the Manchester Ship Canal Railway Department, it would be proper to mention those owned by private firms which either had running rights on the Manchester Ship Canal lines or which handled the internal shunting of the firms connected to the Company's lines.

One of the earliest of these engines is the ex-No. 4 of the Smelting Corporation at Ellesmere Port. In Chapter 2, a description of the locomotive is given together with the fact that it was the only engine traversing the Manchester Ship Canal metals in that section for some time in the early years. This little Borrows well tank is now preserved at the North York Moors Railway. A rather better-known locomotive, which in its time had running rights over the Manchester Ship Canal line, is *Haydock*, an 0-6-0 Stephenson locomotive of 1879. This engine, maker's number 2309, the ex-NCB ex-Richard Evans Haydock Collieries side tank, has been preserved in Penrhyn Castle Museum, Bangor, North Wales.

The Manchester Ship Canal branch, nicknamed 'Queens Road' which ran up to Irlam Station was used solely by the Co-operative Wholesale Society for many years and the CWS had permission to run over the Company's track between the Margarine Works and Queens Road with their 0-4-0 saddle tank locomotives. None of these locomotives are known to have been preserved but one CWS engine that is preserved is the 0-4-0 Peckett fireless locomotive, which was built in 1955, maker's number 2155, and was used for internal shunting at the Margarine Works. After the closure of the CWS internal railway, the engine was presented to Irlam as a children's playpiece and is now sited on a concrete slab in one of the town's play-parks.

Also at Irlam, in the Cadishead area, worked the 0-4-0

Keighley & Worth Valley Railway locomotive No. 31 moves empty rolling stock on a snowy winter day.

R. S. Greenwood

Keighley & Worth Valley Railway locomotive No. 67 drifts down to Haworth with a one coach shuttle on 28th August 1970. The coach was used as the Railway Director's coach in the cinema version of *The Railway Children*. No. 67 also appeared in the opening scene of the film.

John Sagar

Ex-MSC locomotive No. 31, dressed for the role of a Russian tender engine in a BBC Television production of Tolstoy's *Resurrection*. This was not the only role the engine has played whilst at the Keighley & Worth Valley Railway. Other BBC Television appearances were in the *Railway Children* and in a film, *Pit Ponies*.

R. S. Greenwood

Ex-MSC Hudswell Clarke short tank No. 32, *Gothenburg*, seen outside the main building of Bury Transport Museum. After being purchased from the Company in February 1973 by the East Lancs. Light Railway Co., the engine was overhauled and then repainted. The finished colours are light green on tanks, cab and bunker with yellow lining. The buffer beams and side rods are in red with buffers and the smokebox dome in black.

J. Brown

Hudswell Clarke long tank No. 70 stands in the yard at Bury Transport Museum. At the time, it was painted in primer awaiting the finished coats of paint. When completed, the engine was finished in black with red buffer beams, side rods and brake gear.

D. Smith

A selection of locomotive lamps illustrating the changes in style, and size, which took place over the years. The centre lamp fitted on most of the early Hudswell Clarke engines and was still in production into the 1930s. The lamp on the right was only used on a few engines and was still devoid of the side lenses. On the left of the photograph is a rather battered example of the final type of lamp used, and these were fitted with side lenses.

D. Thorpe, Courtesy M S C Co.

Sentinel steam locomotive *Ann*, maker's number 7232. Built in 1927, she worked for The British Tar Products from 1927 to 1971, after which she was withdrawn from service and given to the Yorkshire Dales Railway Society at Embsay on 26th August 1971. Like the CWS fireless locomotive, *Ann* spent most of her working life within the confines of the works premises, simply shunting the wagons to and from the Manchester Ship Canal connection.

Out of the large numbers of rolling stock items, just five have been taken for preservation. These consist of two eight-wheeled coaches, the Mode Wheel breakdown van No. 6374, one wooden box wagon and Cashier's Coach No. 3. In 1972, Dock Labour Board Coach No. 3 was donated to the Historic Rolling Stock Group and was moved to the Severn Valley Railway in June of that year. No. 3 was of special interest to LMS enthusiasts, for the ex-LMS open third No. 7991 is the only example remaining of the 235 which were built in 1926 by the Metropolitan Carriage & Finance Company Ltd. Its former number was 5682. The all-steel coach was to be restored by the new owners to full LMS livery. The second coach to leave the Canal for preservation was the Resident Engineer's Department coach, *Rover*. This vehicle was obtained by the 'Duke of Gloucester Trust', Loughborough, in 1978 for the sum of £500. Built at Wolverton in 1950 to an LMS design, it was given the BR number 27001 and was painted LMS maroon. It was later repainted red and cream and later still this was changed to BR blue. The new owners set to work overhauling the coach, and the intention is to bring it back to the LMS livery.

A more humble item of rolling stock to be saved from the breakers was MSC No. 6115, a wooden-sided ex-GWR box wagon. It is now serving a useful purpose at the Dinting Railway Centre. Cashier's Coach No. 3 went to Steamport, Southport, whilst the ex-GWR 'Toad' was purchased by Rutland Railway Museum, Collismore, near Oakham. It left Mode Wheel on 26th July 1983.

The Manchester Ship Canal Company themselves have retained a few items of railway history, mostly in the form of name and numberplates, but this has been done by individuals in a personal capacity and only occasionally through official instruction.

Before the numbers of steam locomotives had appreciably diminished, several of the brass plates had been sold to members of staff and other interested parties for various reasons and purposes. Some of the numberplates were to be used as house numbers, other plates, including nameplates, have ended up in a place of honour in the homes of ex-drivers who usually bought the plate belonging to the locomotive on which they had spent most of the working hours. A set of plates, including a nameplate, a numberplate, and a maker's plate, were presented, complete with bomb shrapnel scars, to the Manchester Ship Canal representatives in Galveston, that being the name of the locomotive which was No. 16. This set now adorns the wall of the headquarters of the Traffic Club of New York.

It is regrettable that there is so little left which is available for preservation; however, with the items that are still in existence, there is enough to make an interesting exhibition concerning the Manchester Ship Canal Railway. Whether that will ever come about is hard to say but it would be of real interest to a great many ex-Manchester Ship Canal men and railway enthusiasts generally.

Appendix One List of Locomotives

List of Locomotives Working on the Contract

The list has been compiled by members of the Industrial Locomotive Society from various sources including ex-Contract men, and the makers' lists. Locomotives with a question mark alongside the name have not been fully authenticated by the Industrial Locomotive Society.

The list is set out under the makers' names in alphabetical order and not in the order in which they were delivered to the Manchester Ship Canal site.

All dimensions are in feet and inches.

Name	*Makers' No.*	*Type*	*Wheels*	*Cylinder*	*Year*	*Remarks*	
				Black Hawthorn & Co., Gateshead			
Lord Warden	465	0-4-0 or 0-6-0ST	3ft. 2in.	12in. x 18in.	1878	Ex-Barry Dock Contract:	To Staveley Coal & Iron Co.
				Hudswell Clarke & Rodgers			
Wyre	120	0-4-0ST	2ft. 9in.	9in. x 15in.	1872		To J. C. Wills, Heysham
				Hudswell Clarke & Co.			
Flixton	293	0-4-0ST	2ft. 6in.	OC 9in. x 15in.	1887	New to MSC Contract Irlam:	To Sir J. Jackson, Davenport
Thelwall	294	0-4-0ST	2ft. 9in.	OC 10in. x 16in.	1887		To Ammonia Soda Co. Ltd., Plumbley
Bowden	298	0-6-0ST	3ft. 0in.	IC 12in. x 18in.	1888	New to MSC Contract:	To Sir Robt. McAlpine
Stretford	299	0-6-0ST	3ft. 3in.	IC 13in. x 20in.	1888	New to MSC Contract:	To 1890 R. Siddal, Stacksteads Quay
Heatley	300	0-6-0ST	3ft. 3in.	IC 13in. x 20in.	1888	New to MSC Contract:	To Price, Wills & Reeves, Harrow
Baguley	301	0-6-0ST	3ft. 3in.	IC 13in. x 20in.	1888		
Weaste	302	0-4-0ST	2ft. 6in.	OC 9in. x 14in.	1888	New to MSC Contract:	To J. T. Firbank
Hulme	304	0-4-0ST	2ft. 9in.	OC 10in. x 16in.	1888	New to MSC Contract:	
Hyde	306	0-4-0ST	2ft. 6in.	OC 8in. x 15in.	1888	New to MSC Contract:	To Price, Wills & Reeves (Later, Ashington Coal Co. Lancaster)
Lancaster	307	0-6-0T	3ft. 6½in.	IC 15in. x 20in.	1888		
Leigh	310	0-4-0ST	2ft. 9in.	OC 10in. x 16in.	1889		
Medlock	314	0-6-0ST	3ft. 3in.	IC 13in. x 20in.	1888	New to MSC Contract:	To Darwin & Mostyn Coal Co.
Beswick	315	0-6-0ST	3ft. 0in.	IC 12in. x 18in.	1888	New to MSC Contract (Ellesmere Port):	To Topham Jones & Railton (No. 9)
Padgate	316	0-6-0ST	3ft. 0in.	IC 12in. x 18in.	1889	New to MSC Contract:	To Sir Robt. McAlpine
Heaton	318	0-6-0T	3ft. 6½in.	IC 15in. x 20in.	1889	New to MSC Contract:	
Stafford	319	0-6-0T	3ft. 6½in.	IC 15in. x 20in.	1889		to J. C. Wills, Heysham
Worsley	325	0-6-0ST	3ft. 3in.	IC 13in. x 20in.	1889	New to MSC Contract:	To Topham Jones & Railton Crymllyn Burrows (No. 3)
Bromborough	327	0-6-0ST	3ft. 3in.	IC 13in. x 20in.	1889	New to MSC Contract (Ellesmere Port):	To Price, Wills & Reeves
Lord Mayor ?	402	0-4-0ST	2ft. 9in.	OC 10in. x 16in.	1893	New to MSC Contract: (Salford):	Possibly M/cr. Dry Docks
				Hunslet Engine Co.			
Buckland	79	0-4-0ST	2ft. 4in.	8in. x 14in.	1872		To Ravenhead Brick & Tile Co.
Deal	210	0-6-0ST	3ft. 0in.	IC 12in. x 18in.	1879	STC:	To Renishaw Iron Co.
Sale	263	0-4-0ST	3ft. 0in.	OC 13in. x 18in.	1881	Ex-G. B. Godfrey, Hull No.2:	
Greenheys	282	0-4-0ST	3ft. 0in.	OC 12in. x 18in.	1882	Ex-Gibbs & Canning:	To Orrel Colly, Wigan as *Kathleen*
Ashton	336	0-4-0ST	2ft. 9in.	OC 10in. x 15in.	1884	Ex-Preston Dock Contract:	To Leeds Steel Works
Wenvoe	337	0-4-0ST	2ft. 9in.	OC 10in. x 15in.	1884	Ex-TAW (BDC):	To J. Eliss & Son, Barrow-on-Sour *Flossie*
Gaskell	338	0-6-0ST	3ft. 4in.	IC 15in. x 20in.	1884		
Preston	355	0-6-0ST	3ft. 0in.	13in. x 18in.	1884	Ex-PDC:	To R. Holliday & Sons, East Ardsley Coll.
Lytham	360	0-4-0ST	2ft. 9in.	9in. x 14in.	1884	Ex-PDC:	To Sir J. Jackson *Tamar*
Wigan	361	0-4-0ST	2ft. 8in.	9in. x 14in.	1885	New to TAW (Preston):	To John Nelson, then Barnwood Coal Co.
Ribble	362	0-6-0ST	3ft. 0in.	13in. x 17in.	1885	Ex-TAW (PDC):	To Price, Wills & Reeves
Barry	363	0-6-0ST	3ft. 1in.	13in. x 18in.	1885	Ex-BDC:	To Thomas Oliver & Sons
Dynas	368	0-4-0ST	2ft. 8in.	OC 9in. x 14in.	1885	New to TAW (Barry):	
Blackpool	369	0-6-0ST	3ft. 1in.	IC 13in. x 18in.	1885	Ex-PDC:	To Thomas Oliver & Sons
Cardiff	370	0-6-0ST	3ft. 0in.	13in. x 18in.	1885	New to TAW (Preston):	To Hunsbury Hill Furnaces Northampton
Southport	392	0-6-0ST	3ft. 0in.	13in. x 18in.	1886	New to TAW (Preston):	To Price, Wills & Reeves
Monmouth	397	0-6-0T	3ft. 4in.	IC 15½in. x 20in.	1886	New to Preston:	To To Price, Wills & Reeves, Heysham
Gloucester	401	0-6-0T	3ft. 3in.	IC 15in. x 20in.	1886	New to TAW (Barry):	To Sir J. Jackson
Pontypridd	415	0-6-0ST	3ft. 2in.	IC 14in. x 18in.	1887	Ex-BDC:	To Appleby Iron Co. 1894 as No. 5
Wye	420	0-4-0ST	3ft. 2in.	IC 14in. x 18in.	1887	New to TAW (Barry):	
Weston	424	0-4-0ST			1887	New to MSC Contract:	
Ince	425	0-4-0ST			1887	New to MSC Contract:	
Mersey	434	0-6-0ST	3ft. 0in.	IC 13in. x 18in.	1887	New to MSC Contract:	To Thomas Oliver & Sons
Irwell	435	0-6-0ST	3ft. 0in.	IC 14in. x 18in.	1887	New to MSC Contract:	To Garswood Hall Coll. as *Irwell*
Manchester	437	0-6-0ST	3ft. 0in.	IC 14in. x 20in.	1887	New to MSC Contract:	To Sir J. Jackson
Trafford	444	0-6-0ST	3ft. 0in.	13in. x 18in.	1888	New to MSC Contract:	To Pilsley Coll. 1895 as *Trafford*
Pendleton	445	0-6-0ST	3ft. 0in.	13in. x 18in.	1888	New to MSC Contract:	To Appleby Iron Co. 1894 as No. 6

Name	Makers' No.	Type	Wheels	Cylinder	Year	Remarks	
Timperley	449	0-6-0ST	3ft. 0in.	IC 12in. x 18in.	1888	New to MSC Contract:	To Brymbo Steelworks as *Bobs*
Urmston	450	0-6-0ST	3ft. 0in.	IC 12in. x 18in.	1888	New to MSC Contract:	To S. Spearson & Son, Wootton Bassett (c1898)
Lymm	452	0-4-0ST	2ft. 9in.	OC 10in. x 15in.	1888	New to MSC Contract:	Later H. Y. Hey, Blackwoods Quarry, Stacksteads
Dallam	453	0-4-0ST		OC 10in. x ?	1888	New to MSC Contract?:	Later, T. Oliver (Survived after World War II)
Rusholme	454	0-6-0ST	3ft. 0in.	IC 12in. x 18in.	1888	New to MSC Contract:	To H. M. Howell as *Norman*
Bury	456	0-6-0ST	3ft. 1in.	IC 13in. x 18in.	1888	New to MSC Contract:	To Hunbury Hill Furnaces Northampton
Ordsall	457	0-6-0ST	2ft. 11in.	11in. x 16in.	1888	New to MSC Contract:	To Sir J. Jackson *Limerick*
Walton	458	0-4-0ST	2ft. 9in.	10in. x 16in.	1888	New to TAW:	To J. Ellison 1896
Newton	459	0-4-0ST	2ft. 6in.	OC 9in. x 16in.	1889	New to MSC Contract:	Later, Rowntree & Co. York as No. 1
Prescot	468	0-6-0ST	3ft. 0in.	IC 13in. x 18in.	1888	New to MSC Contract:	To Price, Wills & Reeves
Liverpool	469	0-6-0ST	2ft. 9in.	IC 13in. x 18in.	1888	New to MSC Contract:	To Price, Wills & Reeves
Nottingham		0-6-0ST	3ft. 3in.	15in. x 20in.			
Harpurey		0-4-0ST					
Rixton		0-4-0ST					

Kitson & Co.

Name	Makers' No.	Type	Wheels	Cylinder	Year	Remarks	
Stockport	3074	0-6-0T	3ft. 6in.	IC 14in. x 20in.	1888	New to MSC Contract:	To Price, Wills & Reeves as *Stockport*
Chester	3075	0-6-0T	3ft. 6in.	IC 14in. x 20in.	1888	New to MSC Contract:	To WD at Salisbury

Manning Wardle

Name	Makers' No.	Type	Wheels	Cylinder	Year	Remarks	
Swansea	595	0-6-0ST	3ft. 0in.	12in. x 18in.	1876	Ex-Logan & Hemingway:	To Price, Wills & Reeves Heysham
Dover	688	0-6-0ST	3ft. 0in.	IC 12in. x 17in.	1878	New to TAW (Dover):	To A. Dunmore, Three Chimneys Brickworks
River	692	0-4-0ST	3ft. 0in.	IC 10in. x 16in.	1878	New to TAW (Dover):	To Topham Jones & Railton
Walmer	705	0-4-0ST	2ft. 8in.	9in. x 14in.	1878	New to TAW (Dover):	
Martin	714	0-4-0ST	2ft. 10in.	OC 9in. x 14in.	1879	New to TAW (Deal):	To Sir J. Jackson, Davenport, as *Clyde*
Guston	716	0-4-0ST	2ft. 10in.	10in. x 16in.	1878/9	New to TAW (Deal):	To Topham Jones & Railton as *Nell*
Fabian	720	0-4-0ST	2ft. 8in.	OC 9in. x 14in.	1880	New to TAW (Swansea):	To H. Cunliffe, Kettering Brickworks Ltd.
Dungeness	735	0-4-0ST	2ft. 10in.	OC 9in. x 14in.	1881	New to TAW (Appledore):	To George Hargreaves & Co. Boxenden Coll.
Chepstow Caerwent	738	0-6-0ST	3ft. 0in.	12in. x 17in.	1881	New to TAW (Swansea):	To L. P. Nott, Llanelly
Caerwent	785	0-4-0ST	2ft. 10in.	OC 9in. x 14in.	1881	New to TAW (Swansea):	To Dronfield Coll. (also GCR Contract)
Windsor	794	0-4-0ST	2ft. 10in.	9in. x 14in.	1882	New to TAW (Penarth):	
Hafod	838	0-4-0ST		OC 10in. x 18in.	1885	New to Dr. Sillen (Penarth):	To Crossfield & Co., Warrington
Rhondda	839	0-4-0ST		OC 10in. x 18in.	1885	New to Dr. Sillen (Penarth):	To Crossfield & Co., Warrington
Penarth	840	0-6-0ST	2ft. 10in.	IC 13in. x 18in.	1882	Ex-BDC:	To Topham Jones & Railton
Sudbook	891	0-6-0ST	3ft. 0in. or 2ft. 10in.	12in. x 18in. or 12in. x 17in.	1883	New to T. Nelson (Cardiff) No. 16:	To Topham Jones & Railton 1895
Wimborne	933	0-4-0ST	2ft. 10in.	OC 10in. x 16in.	1885	New to TAW (Penarth):	
Accrington	951	0-6-0ST	3ft. 2in.	14in. x 20in.	1885	New to TAW (Ell. Port):	To Sir J. Jackson, Davenport (c1898)
St. Fagans	952	0-4-0ST	3ft. 0in.	12in. x 18in.	1886	Ex-BDC:	To Carrington Saw & Co. St. Helens
General Havelock	957	0-6-0ST		13in. x 18in.	1886	M & N (Brindleheath):	
Cogan	960	0-4-0ST	2ft. 10in.	OC 10in. x 16in.	1885	New to TAW (Penarth):	To Blackpool Corp Gas Co. as *Cogan*
Ferndale	966	0-6-0ST	3ft. 0in.	IC 13in. x 18in.	1888	Ex-BDC:	To Topham Jones & Railton
General Stewart	974	0-6-0ST		IC 12in. x 17in.	1886	M & N (Brindleheath):	To T. Wrigley Prestwich as *Bracken Hill*
Usk	992	0-4-0ST	2ft. 10in.	10in. x 16in.	1887	New to TAW (St. Fagans):	To Arnott Young & Co., Garston
Buckland ?	998	0-4-0ST	2ft. 4in.	8in. x 14in..	1887		
Newport	1001	0-6-0ST	2ft. 10in.	12in. x 17in.	1887	New to TAW (St. Fagans):	To Price & Wills at Barry
St. George(s)	1003	0-6-0ST	3ft. 0in.	12in. x 17in.	1887	New to TAW (St. Fagans):	To T. Oliver & Son, GCR Woodford, Rugby
St. Andrews	1012	0-6-0ST	3ft. 0in.	IC 12in. x 17in.	1887	New to TAW (St. Fagans):	To Topham Jones & Railton
Frodsham	1013	0-6-0ST	2ft. 10in.	IC 12in. x 17in.	1887	New to MSC Contract (Ellesmere Port):	To Topham Jones & Railton as No. 2
Sutton	1016	0-4-0ST		9in. x 14in.	1887	New to MSC Contract (Hooton):	To Topham Jones & Railton
Stanlow	1017	0-4-0ST	2ft. 9in.	9in. x 14in.	1887	New to MSC Contract (Hooton):	To Topham Jones & Railton
Tawe	1020	0-4-0ST	2ft. 10in.	10in. x 16in.	1887	New to TAW (St. Fagans):	To Price, Wills & Reeves
Neath	1021	0-4-0ST		10in. x 16in.	1887	New to TAW (St. Fagans):	To Price, Wills & Reeves
Peterstone	1023	0-4-0ST	3ft. 0in.	OC 12in. x 18in.	1887	New to TAW (St. Fagans):	To Staveley Coal & Iron Co. Staveley
Runcorn	1025	0-6-0ST or 0-6-0T	2ft. 10in.	13in. x 18in.	1887	New to MSC Contract (Runcorn):	To Price, Wills & Reeves, Barry
Hooton	1026	0-4-0ST	3ft. 0in.	12in. x 18in.	1887	New to TAW (Hooton):	
Eastham	1028	0-4-0ST		OC 10in. x 16in.	1887	New to MSC Contract (Eastham):	To C. J. Wills
Barton	1029	0-4-0ST	2ft. 9in.	10in. x 16in.	1887	New to MSC Contract (Ellesmere Port):	To T. A. Ward; also Townsend Watson & Gates
Helsby	1031	0-6-0ST		12in. x 17in.	1887	New to MSC Contract (Hooton):	
Ditton	1033	0-6-0ST	3ft. 0in.	12in. x 18in.	1888	New to MSC Contract (Runcorn):	To Topham Jones & Railton, Newborough

Name	*Makers' No.*	*Type*	*Wheels*	*Cylinder*	*Year*	*Remarks*	
Warburton	1034	0-6-0ST	3ft. 0in.	IC 12in. x 18in.	1888	New to MSC Contract (Latchford):	To Price, Wills & Reeves
Bridgewater	1035	0-6-0ST	3ft. 0in.	13in. x 18in.	1888	New to MSC Contract (Manchester):	To Mole & Newall, GCR, Hawarden Bridge
Barham	1036	0-4-0ST	2ft. 10in.	10in. x 16in.	1888	New to TAW (Barham):	To Sir J. Jackson, Davenport *Torcross 1900*
Irlam	1041	0-4-0ST	not known	12in. x 18in.	1887	New to MSC Contract (Irlam):	To Stringer & Son, Skelmanthorpe, as *Mosley*
Warrington	1047	0-6-0T	3ft. 0in.	13in. x 18in.	1888	New to MSC Contract (Latchford):	To Walter Scott & Middleton Heaton Mersey
Brooklands	1048	0-6-0ST	3ft. 0in.	12in. x 17in.	1888	New to MSC Contract (Latchford):	To T. Oliver & Son, Chapel-en-le-Frith
Ardwick	1040	0-6-0ST	3ft. 0in.	12in. x 17in.	1888	New to MSC Contract (Manchester):	To Price & Wills, Rossyth
Canterbury	1051	0-4-0ST	2ft. 10in.	10in. x 16in.	1888	New to TAW (Barham):	To Sir J. Jackson, Davenport *Tyne*
Moore	1054	0-4-0ST		10in. x 16in.	1888	New to MSC Contract (Latchford):	To Rylands or other wireworks
Hale	1055	0-4-0ST	2ft. 10in.	10in. x 16in.	1888	New to MSC Contract (Latchford):	To C. & W. Walker, Donnington
Brierly	1058	0-6-0ST		IC 12in. x ?	1888	New:	To Ansley Hall Coal & Iron Co. Nuneaton, 1894
Didsbury	1059	0-6-0ST	3ft. 0in.	IC 12in. x 17in.	1888	New to TAW (Irlam):	To Walker, Scott & Middleton as *Berthen*
Norton	1061	0-4-0ST	2ft. 10in.	12in. x 18in.	1888	New to MSC Contract (Irlam):	To Whitecross Co., Warrington
Broughton	1062	0-6-0T	3ft. 0in.	13in. x 18in.	1888	New to MSC Contract (Ellesmere Port):	
Cornbrook	1063	0-6-0T	3ft. 0in.	IC 13in. x 18in.	1888	New to MSC Contract (Patricroft):	To C. J. Wills, then Price, Wills & Reeves
Oldham	1064	0-6-0ST	3ft. 0in.	13in. x 18in.	1888	New to MSC Contract (Latchford):	
Rochdale	1065	0-6-0T	3ft. 0in.	13in. x 18in.	1888	New to MSC Contract (Ellesmere Port):	
Clitheroe	1071	0-6-0ST	3ft. 0in.	IC 12in. x 17in.	1888	New to MSC Contract (Latchford):	To J. Jackson, later Pauling & Co.
Weaver	1072	0-4-0ST	2ft. 10in.	OC 12in. x 18in. or OC 10in. x 16in.	1888	New to MSC Contract (Manchester):	To W. Brown & Sons, Contr. N. Staffs. 1898
Seedley	1080	0-4-0ST		10in. x 16in.	1888	New to MSC Contract (Latchford):	
Carrington	1081	0-4-0ST	2ft. 9in.	OC 10in. x 16in.	1887	New to MSC Contract (Irlam):	To J. H. Lawson & Co., Neath (S. Pearson & Son)
Fleetwood	1084	0-4-0ST		12in. x 18in.	1888	New to TAW (Preston):	To N. Greening & Son, Warrington. 1901
Chorley	1085	0-4-0ST	2ft. 11in.	OC 12in. x 18in.	1888	New to TAW (Preston):	To Price & Wills, Barry Docks
Sankey	1088	0-4-0ST	2ft. 10in.	OC 10in. x 16in.	1888	New to MSC Contract (Ellesmere Port):	To Topham Jones & Railton No. No. 12 C.B.
Disley	1089	0-4-0ST	2ft. 10in.	OC 10in. x 16in.	1888	New to MSC Contract (Latchford):	To Sir John Jackson *Warrington*
Cadishead	1090	0-4-0ST	3ft. 0in.	OC 12in. x 18in.	1888	New to MSC Contract (Irlam):	To Sir J. Jackson, Davenport *Lord Nelson*
Ancoats	1091	0-4-0ST	3ft. 0in.	12in. x 18in.	1888	New to MSC Contract (Ellesmere Port):	
Blackburn	1099	0-6-0T	3ft. 6in.	14in. x 20in.	1888	New to MSC Contract (Manchester):	To Naylor Bros. Huddersfield
Birkenhead	1100	0-6-0ST	3ft. 0in.	13in. x 18in.	1888	New to MSC Contract (Irlam):	To Price, Wills & Reeves, BDC
Darwen	1101	0-6-0ST	3ft. 0in.	IC 13in. x 18in.	1888	New to MSC Contract (Ellesmere Port):	To L. P. Nott, Llanelly as *Llanelly*
Prestwich	1102	0-6-0ST	3ft. 0in.	IC 13in. x 18in.	1889	New to MSC Contract (Salford):	To Griff Colly., Nuneaton, *Bob*
Levenshulme	1103	0-6-0ST	3ft. 0in.	13in. x 18in.	1889	New to MSC Contract (Ellesmere Port):	to Topham Jones & Railton *No. 3 Crymllyn Burrows*
Gowy	1119	0-6-0ST	3ft. 1in.	IC 12in. x 17in.	1889	New to MSC Contract (Cornbrook):	To Topham Jones & Railton
Brough ?		0-4-0ST				Alleged to have collided with *Bromborough*	
Hardwicke ?		0-6-0ST		IC		Eye-witness only	
Huyton ?		0-4-0ST				Eye-witness only	
Marlow ?		0-4-0ST				Eye-witness only	
St. Annes ?		0-6-0ST				Eye-witness only	
St. Helens		0-6-0T				Eye-witness only	

Peckett & Sons

Name	*Makers' No.*	*Type*	*Wheels*	*Cylinder*	*Year*	*Remarks*	
Ellesmere	456	0-6-0ST	3ft. 6in.	12in. x 18in.	1887	New to MSC Contract (Ellesmere Port):	To Anglo-American Oil Co. as *Jack*

Ruston & Proctor

Name	*Makers' No.*	*Type*	*Wheels*	*Cylinder*	*Year*	*Remarks*	
Eccles	13003	0-6-0T	3ft. 0in.	12in. x 18in.	1888	New to MSC Contract:	To Monk & Newall, LNWR (Tamworth)
Widnes	13111	0-4-0ST	2ft. 8in.	10in. x 16in.	1888	New to MSC Contract:	To Bowman Thompson, Lostock
Elton	13112	0-4-0ST	2ft. 8in.	10in. x 16in.	1888	New to MSC Contract:	To Constable Hart & Co. Matlock
Glazebrook	13113	0-4-0ST	2ft. 8in.	OC 10in. x 16in.	1888	New to MSC Contract:	To Price, Wills & Reeves
Lincoln		0-6-0T				Eye-witness only	

Name	Makers' No.	Type	Wheels	Cylinder	Year	Remarks	
				Sharp Stewart & Co.			
Haro Haro	2047	2-4-0T	4ft. 0in.	IC 10in. x 16in.	1870	Ex-Jersey Railway:	
Duke of Normandy	2048	2-4-0T	4ft. 0in.	IC 10in. x 16in.	1870	Ex-B DC:	To Vulcan Foundry, via T. W. Ward, 1897
Knutsford	3471	0-6-0T	3ft. 0in.	13in. x 20in.	1888	New to MSC Contract:	To Broughton & Plas Power Coal Co.
Tatton	3472	0-6-0T	3ft. 0in.	13in. x 20in.	1888	New to MSC Contract:	To Sir J. Jackson, later L. P. Nott, as *Sharpness*
Openshaw	3474	0-6-0T	3ft. 0in.	13in. x 20in.	1888	New to MSC Contract	To Sir J. Jackson, Davenport *148 Westminster*
Stalybridge	3475	0-6-0ST or 0-6-0T	3ft. 0in.	13in. x 20in.	1888	New to MSC Contract	To Aston Hall Colly Col., Flintshire
Dunham ?		0-6-0T				Eye-witness only	
				Vulcan Engine Co.			
Pemberton	798	0-4-0ST		OC	1876		
Patricroft	1232	0-6-0ST	3ft. 1in.	12in. x 17in.	1888	New to MSC Contract:	To Price, Wills & Reeves
Kenyon	1233	0-6-0ST	2ft. 11in.	IC 12in. x 17in.	1888	New to MSC Contract:	To Price, Wills & Reeves
Daresbury	1234	0-6-0ST	3ft. 1in.	12in. x 17in.	1888	New to MSC Contract:	To Price, Wills & Reeves
Northenden	1235	0-6-0ST	3ft. 1in.	12in. x 17in.	1888	New to MSC Contract:	To Price, Wills & Reeves
Partington	1236	0-6-0ST	3ft. 0in.	12in. x 17in.	1888	New to MSC Contract:	To Price, Wills & Reeves
				Walker Bros.			
Bradford		0-6-0T	3ft. 0in.	IC 13in. x 20in.	1888	New to MSC Contract:	
Burnley		0-6-0T	3ft. 0in.	IC 13in. x 20in.	1888	New to MSC Contract:	

Locomotives retained from Contract for Manchester Ship Canal Railway purposes

Name	Makers' No.	Type	Wheels	Cylinder	Year	Remarks	
				Hudswell Clarke & Co.			
Cheadle	305	0-4-0ST	2ft. 9in.	OC 10in. x 16in.	1888	New to MSC Contract (Cornbrook):	Sold Pre-1914
				Hunslet Engine Co.			
Rhymney	380	0-4-0ST	3ft. 0in.	12in. x 18in.	1886	Ex-BDC to MSC Contract:	Sold Pre-1914
Glamorgan	396	0-6-0T	3ft. 3in.	IC 15in. x 20in.	1886	Ex-BDC to MSC Contract:	Sold Pre-1914
Bollin	462	0-6-0ST	3ft. 1in.	IC 12in. x 18in.	1889	New to MSC Contract:	Sold Pre-1914
Fallowfield	463	0-6-0ST	3ft. 1in.	IC 12in. x 18in.	1888	New to MSC Contract:	Sold 1913/4
				Manning Wardle & Co.			
Severn	746	0-6-0ST	3ft. 0in.	IC 15in. x 22in.	1880	To TAW *Deal* to MSC Contract:	Sold 1910
Salford	1024	0-6-0T or 0-6-0ST		13in. x 18in.	1888	New *Salford* to MSC Contract:	Sold Pre-1914
Grange	1027	0-4-0ST	3ft. 0in.	OC 12in. x 18in.	1887	New *Ince* to MSC Contract:	Sold to J. Crossfield as *Warrington*
Latchford	1030	0-6-0ST		12in. x 17in.	1887	New *Ellesmere Port* to MSC Contract:	Sold Pre-1914
Chorlton	1060	0-4-0ST	3ft. 0in.	12in. x 18in.	1888	New *Manchester* to MSC Contract:	Sold Pre-1914
				Ruston & Proctor			
Altrincham	13044	0-6-0T	3ft. 4in.	12in. x 18in.		New to MSC Contract:	
				Sharp Stewart & Co.			
Northwich	3473	0-6-0T	3ft. 0in.	IC 13in. x 20in.	1889	New to MSC Contract:	Sold Pre-1914
				Walker Bros.			
Withington		0-6-0T	3ft. 0in.	IC 13in. x 20in.	1888		

No.	Name	Maker's Name	Maker's No.	Type	Wheels		Cylinders	Year	Remarks
1	*Elham*	Hunslet Engine Co.	353	0-6-0ST	3ft. 3in.	IC	12in. x 20in.	1884	Scrapped 1929 or 1933
2	*Romilly*	Hunslet Engine Co.	357	0-6-0ST	3ft. 0in.	IC	12in. x 20in.	1885	Sold 1932
3	*Raglan*	Hunslet Engine Co.	414	0-4-0ST			13in. x 18in.	1887	Sold
4	*Cadoxton*	Hunslet Engine Co.	393	0-6-0ST	3ft. 1in.	IC	13in. x 18in.	1886	Sold
5	*Douglas*	Hunslet Engine Co.	381	0-4-0ST	3ft. 1in.	OC	13in. x 18in.	1886	Sold
6	*Longsight*	Hunslet Engine Co.	455	0-6-0ST	3ft. 0in.	OC	12in. x 18in.	1888	Sold 1931
7	*Bolton*	Manning Wardle & Co.	1066	0-6-0ST	3ft. 3in.	IC	12in. x 17in.	1888	Sold March 1905
8	*Heywood*	Manning Wardle & Co.	1082	0-6-0ST	3ft. 6in.	IC	14in. x 20in.	1888	Sold 1927
9	*Winsford*	Hudswell Clarke & Co.	348	0-6-0ST	3ft. 3in.	IC	13in. x 20in.	1892	Sold 1922
10	*Irk*	Hunslet Engine Co.	566	0-6-0ST	3ft. 3in.	IC	15in. x 20in.	1892	Sold 1927

No.	Name	Maker's Name	Maker's No.	Type	Wheels		Cylinders	Year	Remarks
		Locomotives Purchased Specifically for Manchester Ship Canal Railway							
11	*Alexandria*	Peckett & Sons	654	0-4-0ST	3ft. 3½in.	OC	14in. x 20in.	1897	Sold to Esso (Trafford Wharf) Mcr. August 1954
12	*Jaffa*	Peckett & Sons	655	0-4-0ST		OC	14in. x 20in.	1897	Sold 1927
13	*Montreal*	Hunslet Engine Co.	685	0-6-0T	3ft. 4in.	IC	15½in. x 20in.	1898	Scrapped 1957
14	*St. John*	Hunslet Engine Co.	686	0-6-0T		IC	15½in. x 20in.	1898	Sold to ICI, Blackley 1st May 1963
15	*New York*	Hunslet Engine Co.	699	0-6-0T		IC	15½in. x 20in.	1899	Scrapped June 1963
16	*Galveston*	Hunslet Engine Co.	700	0-6-0T		IC	15½in. x 20in.	1899	Scrapped June 1963
17	*New Orleans*	Hunslet Engine Co.	711	0-6-0T		IC	15½in. x 20in.	1899	Scrapped January 1956
18	*Savannah*	Hunslet Engine Co.	712	0-6-0T		IC	15½in. x 20in.	1899	Scrapped August 1957
19	*London*	Hunslet Engine Co.	721	0-6-0T		IC	15½in. x 20in.	1900	Scrapped June 1963
20	*Glasgow*	Hunslet Engine Co.	732	0-6-0T		IC	15½in. x 20in.	1900	Scrapped June 1963
21	*Antwerp*	Hunslet Engine Co.	777	0-6-0T		IC	15½in. x 20in.	1902	Scrapped June 1963
22	*Rotterdam*	Hunslet Engine Co.	778	0-6-0T		IC	15½in. x 20in.	1902	Scrapped May 1966
23	*Boston*	Hudswell Clarke & Co.	641	0-6-0T	3ft. 4in.	IC	15½in. x 20in.	1902	Scrapped June 1963
24	*Philadelphia*	Hudswell Clarke & Co.	642	0-6-0T	3ft. 4in.	IC	15½in. x 20in.	1902	Scrapped December 1964
25	*Valencia*	Peckett & Sons	907	0-4-0ST	3ft. 3½in.	IC	14in. x 20in.	1902	Scrapped 1957
26	*Bristol*	Hudswell Clarke & Co.	615	0-4-0ST	3ft. 1in.	OC	13in. x 18in.	1903	Sold to Lancs. Steel Corporation 1938
27	*Dublin*	Hudswell Clarke & Co.	643	0-6-0T	3ft. 1in.	IC	15½in. x 20in.	1903	Scrapped June 1963
28	*Belfast*	Hudswell Clarke & Co.	644	0-6-0T	3ft. 1in.	IC	15½in. x 20in.	1903	Scrapped June 1963
29	*Bombay*	Hudswell Clarke & Co.	662	0-6-0T	3ft. 1in.	IC	15½in. x 20in.	1903	Scrapped 5th May 1966
30	*Kurrachee*	Hudswell Clarke & Co.	663	0-6-0T	3ft. 1in.	IC	15½in. x 20in.	1903	Scrapped 5th October 1977
31	*Hamburg*	Hudswell Clarke & Co.	679	0-6-0T	3ft. 1in.	IC	15½in. x 20in.	1903	Sold to Mr Greenwood, Keighley & Worth Valley Railway, June 1976
32	*Gothenburg*	Hudswell Clarke & Co.	680	0-6-0T	3ft. 1in.	IC	15½in. x 20in.	1903	Sold to East Lancs. Light Railway Co. (Bury) February 1973
33	*Christiania*	Hudswell Clarke & Co.	695	0-6-0T	3ft. 1in.	IC	15½in. x 20in.	1904	Scrapped 1959
34	*Copenhagen*	Hudswell Clarke & Rodgers	218	0-6-0ST	3ft. 8½in.	OC	14in. x 20in.	1880	Sold to W. Arnott Young & Co. 1938
34	*Jeannie*	Hudswell Clarke & Co.	694	0-6-0T	3ft. 1in.	IC	15½in. x 20in.	1904	Scrapped 23rd August 1958
35	*Stockholm*	Hudswell Clarke & Co.	701	0-6-0T	3ft. 1in.	IC	15½in. x 20in.	1904	Scrapped May 1966
36	*Port Limon*	Kitson & Co.	4295	0-6-0T	4ft. 6in.	IC	18in x 26in.,	1904	Scrapped January 1954
37	*Adelaide*	Hudswell Clarke & Co.	735	0-6-0T	3ft. 1in.	IC	15½in. x 20in.	1905	Scrapped May 1964
38	*Melbourne*	Hudswell Clarke & Co.	739	0-6-0T	3ft. 1in.	IC	15½in. x 20in.	1905	Scrapped June 1962
39	*Sydney*	Hudswell Clarke & Co.	740	0-6-0T	3ft. 1in.	IC	15½in. x 20in.	1905	Scrapped May 1966
40	*Buenos Ayres*	Hudswell Clarke & Co.	755	0-6-0T	3ft. 1in.	IC	15½in. x 20in.	1906	Scrapped 1959
41	*Monte Video*	Hudswell Clarke & Co.	756	0-6-0T	3ft. 1in.	IC	15½in. x 20in.	1906	Scrapped 1959
42	*Rosario*	Hudswell Clarke & Co.	757	0-6-0T	3ft. 1in.	IC	15½in. x 20in.	1906	Scrapped May 1964
43	*Wellington*	Hudswell Clarke & Co.	758	0-6-0T	3ft. 1in.	IC	15½in. x 20in.	1906	Scrapped 1959
44	*Auckland*	Hudswell Clarke & Co.	775	0-6-0T	3ft. 1in.	IC	15½in. x 20in.	1906	Scrapped October 1967
45	*Jamaica*	Hudswell Clarke & Co.	780	0-6-0T	3ft. 1in.	IC	15½in. x 20in.	1906	Scrapped 1959
46	*Quebec*	Hudswell Clarke & Co.	781	0-6-0T	3ft. 1in.	IC	15½in. x 20in.	1907	Scrapped May 1964
47	*Canada*	Hudswell Clarke & Co.	789	0-6-0T	3ft. 1in.	IC	15½in. x 20in.	1907	Scrapped October 1961
48	*India*	Hudswell Clarke & Co.	790	0-6-0T	3ft. 1in.	IC	15½in. x 20in.	1907	Scrapped 1958
49	*Stettin*	Hudswell Clarke & Co.	797	0-6-0T	3ft. 1in.	IC	15½in. x 20in.	1907	Scrapped March 1964
50	*Australia*	Hudswell Clarke & Co.	798	0-6-0T	3ft. 1in.	IC	15½in. x 20in.	1907	Scrapped October 1961
51	*New Zealand*	Hudswell Clarke & Co.	799	0-6-0T	3ft. 1in.	IC	15½in. x 20in.	1907	Scrapped May 1966
52	*Norway*	Hudswell Clarke & Co.	940	0-6-0T	3ft. 1in.	IC	15½in. x 20in.	1911	Scrapped October 1967
53	*Sweden*	Hudswell Clarke & Co.	941	0-6-0T	3ft. 1in.	IC	15½in. x 20in.	1911	Scrapped May 1966
54	*Odessa*	Hudswell Clarke & Co.	1001	0-6-0T	3ft. 1in.	IC	15½in. x 20in.	1912	Scrapped May 1966
55	*Batoum*	Hudswell Clarke & Co.	1002	0-6-0T	3ft. 1in.	IC	15½in. x 20in.	1912	Scrapped May 1964
56	*Smyrna*	Hudswell Clarke & Co.	1020	0-6-0T	3ft. 1in.	IC	15½in. x 20in.	1913	Scrapped June 1963
57	*Dunkirk*	Hudswell Clarke & Co.	1041	0-6-0T	3ft. 1in.	IC	15½in. x 20in.	1913	Scrapped May 1966
58	*Bordeaux*	Hudswell Clarke & Co.	1042	0-6-0T	3ft. 1in.	IC	15½in. x 20in.	1913	Scrapped October 1967
59	*Ghent*	Hudswell Clarke & Co.	1046	0-6-0T	3ft. 1in.	IC	15½in. x 20in.	1913	Scrapped May 1966
60	*Riga*	Hudswell Clarke & Co.	1047	0-6-0T	3ft. 1in.	IC	15½in. x 20in.	1913	Scrapped June 1964
61	*Reval*	Hudswell Clarke & Co.	1059	0-6-0T	3ft. 1in.	IC	15½in. x 20in.	1914	Scrapped May 1966
62	*St. Petersburg*	Hudswell Clarke & Co.	1060	0-6-0T	3ft. 1in.	IC	15½in. x 20in.	1914	Scrapped October 1967
63		Hudswell Clarke & Co.	1224	0-6-0T	3ft. 1in.	IC	15½in. x 20in.	1916	Scrapped October 1967
64		Hudswell Clarke & Co.	1240	0-6-0T	3ft. 1in.	IC	15½in. x 20in.	1916	Scrapped October 1967
65		Hudswell Clarke & Co.	1361	0-6-0T	3ft. 1in.	IC	15½in. x 20in.	1919	Scrapped October 1967
66		Hudswell Clarke & Co.	1362	0-6-0T	3ft. 1in.	IC	15½in. x 20in.	1919	Scrapped June 1963
67		Hudswell Clarke & Co.	1369	0-6-0T	3ft. 1in.	IC	15½in. x 20in.	1919	Sold to Mr Greenwood, Keighley & Worth Valley Railway, June 1967
68		Hudswell Clarke & Co.	1370	0-6-0T	3ft. 1in.	IC	15½in. x 20in.	1919	Scrapped October 1967
69		Hudswell Clarke & Co.	1430	0-6-0T	3ft. 1in.	IC	15½in. x 20in.	1921	Scrapped June 1964
70		Hudswell Clarke & Co.	1464	0-6-0T	3ft. 1in.	IC	15½in. x 20in.	1921	Sold to East Lancs Light Rly (Bury) June 1973
71		Hudswell Clarke & Co.	1465	0-6-0T	3ft. 1in.	IC	15½in. x 20in.	1921	Scrapped June 1964
72		Kitson & Co.	5357	0-6-0T	4ft. 3in.	IC	18in. x 26in.	1921	Scrapped December 1964
73		Kitson & Co.	5370	0-6-0T	4ft. 3in.	IC	18in. x 26in.	1923	Scrapped March 1964
74		Kitson & Co.	5371	0-6-0T	4ft. 3in.	IC	18in. x 26in.	1923	Scrapped June 1963
75		Hudswell Clarke & Co.	1527	0-6-0ST	3ft. 1in.	IC	12in. x 18in.	1924	Scrapped 1953
76		Hudswell Clarke & Co.	1589	0-6-0T	3ft. 1in.	IC	15½in. x 20in.	1927	Scrapped December 1964
77		Peckett & Sons	1724	0-4-0ST	3ft. 4in.	OC	14in. x 22in.	1927	Scrapped 13th June 1959
78		Hudswell Clarke & Co.	1584	0-6-0ST	3ft. 1in.	IC	12in. x 18in.	1927	Scrapped January 1962
79		Hudswell Clarke & Co.	1590	0-6-0T	3ft. 1in.	IC	15½in. x 20in.	1927	Scrapped October 1967
80		Kitson & Co.	5321	0-6-0T	4ft. 3in.	IC	18in. x 26in.	1927	Scrapped December 1964
81		Kitson & Co.	5422	0-6-0T	4ft. 3in.	IC	18in. x 26in.	1929	Scrapped March 1964
82		Kitson & Co.	5423	0-6-0T	4ft. 3in.	IC	18in. x 26in.	1929	Scrapped March 1964
83		Hunslet Engine Co.	1694	0-6-0ST	3ft. 0in.	IC	12in. x 18in.	1932	Scrapped June 1963
84		Hunslet Engine Co.	1695	0-6-0ST	3ft. 0in.	IC	12in. x 18in.	1932	Scrapped December 1957
85	(ex-WD75304)	Vulcan Engine Co.	5294	0-6-0ST	4ft. 3in.	IC	18in. x 26in.	1945	Scrapped March 1964
86	(ex-WD71493)	Hudswell Clarke & Co.	1769	0-6-0ST	4ft. 3in.	IC	18in. x 26in.	1944	Scrapped March 1964
87	(ex-WD71441)	Hunslet Engine Co.	3205	0-6-0ST	4ft. 3in.	IC	18in. x 26in.	1945	Scrapped March 1964
88		Hudswell Clarke & Co.	1865	0-6-0ST	4ft. 3½in.	OC	18in. x 26in.	1954	Scrapped March 1964
89		Hudswell Clarke & Co.	1866	0-6-0ST	4ft. 3½in.	OC	18in. x 26in.	1954	Scrapped December 1964
90		Hudswell Clarke & Co.	1867	0-6-0ST	4ft. 3½in.	OC	18in. x 26in.	1954	Scrapped December 1964

Diesel Locomotives

MSC No.	*Name*	*Maker*	*Works No.*	*Wheels*	*Year*	*Remarks*
4001	*Alnwick Castle*	Hudswell Clarke & Co.	D1075	0-6-0	1959	
4002	*Arundel Castle*	Hudswell Clarke & Co.	D1076	0-6-0	1959	
2001 (D1)	*Ashdown*	Hudswell Clarke & Co.	D1186	0-6-0	1959	Sold to Hunslet (Holdings) Ltd., Leeds, April 1976
D2		Hudswell Clarke & Co.	D1187	0-6-0	1960	
D3		Hudswell Clarke & Co.	D1188	0-6-0	1960	
D4		Hudswell Clarke & Co.	D1189	0-6-0	1960	Sold to Andrew Barclay, February 1976
D5		Hudswell Clarke & Co.	D1190	0-6-0	1960	Withdrawn June 1983
D6		Hudswell Clarke & Co.	D1191	0-6-0	1960	
D7		Hudswell Clarke & Co.	D1251	0-6-0	1962	Withdrawn June 1983
D8		Hudswell Clarke & Co.	D1252	0-6-0	1962	Withdrawn June 1983
D9		Hudswell Clarke & Co.	D1253	0-6-0	1962	Sold to Hunslet (Holdings) Ltd., Leeds, January 1976
D10		Hudswell Clarke & Co.	D1254	0-6-0	1962	Sold to Hunslet (Holdings) Ltd., Leeds, January 1976
D11		Hudswell Clarke & Co.	D1255	0-6-0	1962	
D12		Hudswell Clarke & Co.	D1256	0-6-0	1962	Withdrawn June 1983, dismantled
D13		Hudswell Clarke & Co.	D1257	0-6-0	1962	Withdrawn June, 1983, dismantled
D14		Hudswell Clarke & Co.	D1258	0-6-0	1962	Withdrawn 13th Nov. 1983 and dismantled
E1		Hudswell Clarke & Co.	D1199	0-6-0	1960	Stored at Mode Wheel
DH3001		Rolls-Royce (Sentinel)	10144	0-6-0	1963	Stationed at Ellesmere Port/Stanlow
DH3002		Rolls-Royce (Sentinel)	10145	0-6-0	1963	Stationed at Ellesmere Port/Stanlow
DH3003		Rolls-Royce (Sentinel)	10146	0-6-0	1963	Stationed at Ellesmere Port/Stanlow
DH3004		Rolls-Royce (Sentinel)	10147	0-6-0	1963	Stationed at Ellesmere Port/Stanlow
DH3005		Rolls-Royce (Sentinel)	10162	0-6-0	1963	Stationed at Ellesmere Port/Stanlow
DH15		Rolls-Royce (Sentinel)	10174	0-4-0	1963	Sold to Bowater Ltd., (Ellesmere Port) November 1970
DH16		Rolls-Royce (Sentinel)	10175	0-4-0	1964	Sold to Bowater Ltd., (Ellesmere Port) November 1970
DH17		Rolls-Royce (Sentinel)	10176	0-4-0	1964	Sold to Thomas Hill (Rotherham) Ltd., March 1971
DH18		Rolls-Royce (Sentinel)	10177	0-4-0	1964	Sold to Thomas Hill (Rotherham) Ltd., July 1972
DH19		Rolls-Royce (Sentinel)	10193	0-4-0	1964	Sold to Thomas Hill (Rotherham) Ltd., April 1972
DH20		Rolls-Royce (Sentinel)	10194	0-4-0	1964	Sold to Thomas Hill (Rotherham) Ltd., November 1971
DH21		Rolls-Royce (Sentinel)	10195	0-4-0	1964	Sold to Thomas Hill (Rotherham) Ltd., December 1971
DH22		Rolls-Royce (Sentinel)	10196	0-4-0	1964	Stripped for spares, December 1971
DH23		Rolls-Royce (Sentinel)	10226	0-4-0	1965	Modified for tandem working between Dec. 1971 and 2nd
DH24		Rolls-Royce (Sentinel)	10227	0-4-0	1965	January 1972
DH25		Rolls-Royce (Sentinel)	10228	0-4-0	1965	Sold to Thomas Hill (Rotherham) Ltd., August 1971
DH26		Rolls-Royce (Sentinel)	10229	0-4-0	1965	Modified for tandem working Dec. 1971 to Jan. 1972
DH27		Rolls-Royce (Sentinel)	10241	0-4-0	1966	Sold to Thomas Hill (Rotherham) Ltd., November 1971
DH28		Rolls-Royce (Sentinel)	10242	0-4-0	1966	Sold to Thomas Hill (Rotherham) Ltd., March 1972
DH29		Rolls-Royce (Sentinel)	10243	0-4-0	1966	Sold to British Salt Limited, February 1972
DH30		Rolls-Royce (Sentinel)	10244	0-4-0	1966	Sold to Thomas Hill (Rotherham) Ltd., March 1972
DH31		Rolls-Royce (Sentinel)	10245	0-4-0	1966	Sold to Thomas Hill (Rotherham) Ltd., January 1972
DH32		Rolls-Royce (Sentinel)	10246	0-4-0	1966	Sold to Thomas Hill (Rotherham) Ltd., February 1972

Appendix Two — Tonnage Totals

Year	Shipping (Tons)	Railway (Tons)
1894	925,659	45,000
1895	1,358,875	550,000
1896	1,826,237	600,000
1897	2,065,815	900,000
1898	2,595,585	1,260,000
1899	2,778,108	1,370,000
1900	3,060,516	1,750,000
1901	2,942,393	1,550,000
1902	3,418,059	
1903	3,846,895	
1904	3,917,578	
1905	4,253,354	
1906	4,700,924	
1907	5,210,759	
1908	4,582,496	
1909	4,563,401	
1910	4,937,631	
1911	5,217,812	
1912	5,339,884	
1913	5,780,161	3,779,757
1914	5,424,732	4,208,684
1915	5,434,046	4,985,940
1916	4,858,749	5,212,702
1917	4,152,754	5,593,635
1918	3,497,995	5,387,592
1919	3,589,043	4,333,397
1920	4,387,363	4,948,169
1921	3,293,670	3,421,668
1922	4,273,544	4,560,663
1923	5,363,940	5,938,197
1924	5,435,508	5,575,969
1925	5,881,691	5,460,699
1926	6,830,879	4,957,044
1927	6,259,420	5,443,422
1928	6,256,895	5,252,166
1929	6,558,589	5,781,409
1930	6,290,625	5,454,849
1931	5,898,289	5,269,397
1932	5,318,935	4,828,569
1933	5,389,069	5,237,349
1934	5,949,970	5,683,523
1935	6,135,003	5,626,430
1936	6,373,308	5,992,935
1937	7,053,216	6,614,652
1938	6,409,873	5,912,162

Year	Shipping (Tons)	Railway (Tons)
1939	7,193,330	6,462,014
1940	8,116,063	7,773,535
1941	7,938,118	8,074,951
1942	8,110,534	8,686,419
1943	7,532,467	8,470,534
1944	7,738,754	8,463,312
1945	6,531,963	7,325,824
1946	5,699,125	5,714,855
1947	6,959,988	6,190,986
1948	8,248,116	6,770,669
1949	9,035,521	7,449,172
1950	9,886,322	7,151,905
1951	10,948,206	6,940,302
1952	11,861,449	6,997,152
1953	12,485,071	7,004,842
1954	16,359,612	6,990,089
1955	18,563,376	6,953,074
1956	18,505,062	6,782,625
1957	16,838,510	5,713,029
1958	17,984,053	4,160,148
1959	18,558,210	4,011,654
1960	17,707,422	4,495,134
1961	14,900,656	4,046,677
1962	14,737,434	3,519,624
1963	15,112,884	3,712,675
1964	15,693,399	3,867,374
1965	15,715,409	3,580,092
1966	16,724,561	3,769,390
1967	16,824,116	3,123,224
1968	16,451,318	2,956,873
1969	16,484,764	2,250,035
1970	15,949,697	2,060,339
1971	17,124,212	1,821,960
1972	16,024,288	1,499,662
1973	16,420,499	1,386,268
1974	17,369,123	1,152,936
1975	14,816,121	1,293,094
1976	15,710,338	1,311,923
1977	14,505,973	1,272,571
1978	14,909,136	1,134,509
1979	13,742,400	1,224,253
	(Tonnes)	*(Tonnes)*
1980	12,503,234	1,120,813
1981	11,738,362	825,910
1982	12,293,637	818,749
1983	11,631,217	886,937

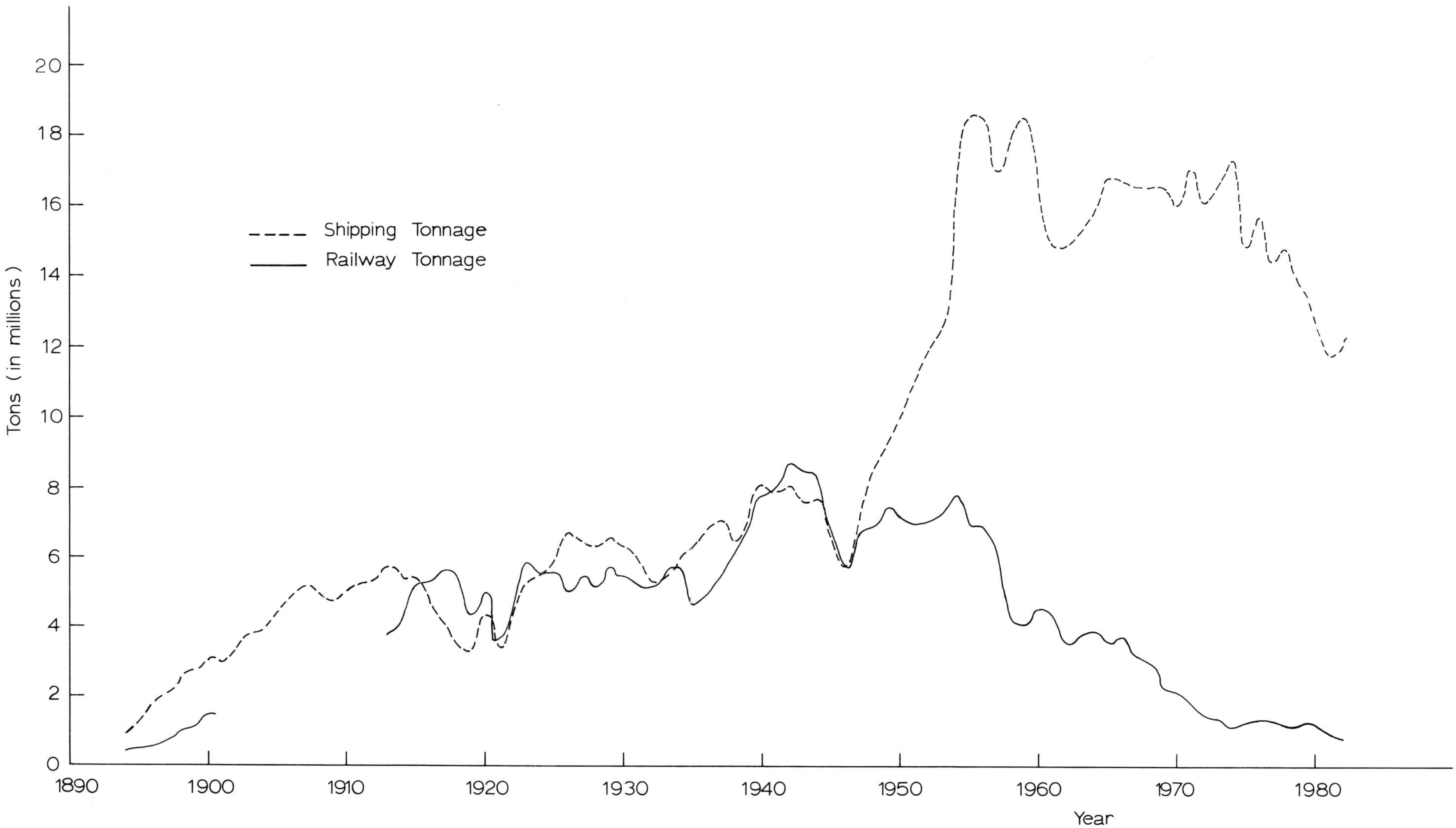

Graph of railway and shipping tonnage carried since 1894.

Donald Thorpe

Appendix Three

List of Officials (1894-1984)

Chief Engineers

E. L. Williams	1887-1895
W. H. Hunter	1895
H. Congreve	
H. A. Reed	
F. B. Greenwood	
R. D. Brown	
D. C. Milne	
O. Hardie	
A. E. Blyth	
S. Howarth	

Railway Superintendents

H. H. Hamilton	1894-1920
A. Alcock	1920-1933
C. E. Hutchinson	1933-1947
W. Heywood	1947-1957
G. A. Jones	1957-1966
W. H. Croft	1966-1968
B. Valentine	1968-Feb. 1982*

Note: Position terminated February 1982. Responsibility for Railway Department taken over by Assistant Manager (Developments) February 1982, J. B. Chilton.

Chief Mechanical and Electrical Engineers

W. Fox	1894-1926
W. G. Smith	1926-1941
T. A. Guest	1941-1951
C. H. Gibson	1951-1962
A. S. J. Campbell	1962-1982
J. Coupe	1982-

Locomotive Superintendents

W. Lewis	-1935
J. Threadgold	1935-1956
A. Blanchard	1956-1962
G. Tudge	1962-1965
J. Steggles	1965-1974
Position terminated	13/6/1974